Lecture Notes in Earth Sciences

Lecture Notes in Earth Sciences

Edited by Somdev Bhattacharji, Gerald M. Friedman,
Horst J. Neugebauer and Adolf Seilacher

23

Karl B. Föllmi

Evolution of
the Mid-Cretaceous Triad

Platform Carbonates, Phosphatic Sediments, and Pelagic
Carbonates Along the Northern Tethys Margin

Springer-Verlag Berlin Heidelberg GmbH

Author

Dr. Karl B. Föllmi
Earth Sciences Board, University of California
Santa Cruz, CA 95064, USA

from Oct. 1989:
Geological Institute, ETH-Centre
CH-8092 Zürich, Switzerland

ISBN 978-3-540-51359-9   ISBN 978-3-540-46199-9 (eBook)
DOI 10.1007/978-3-540-46199-9

© Springer-Verlag Berlin Heidelberg 1989

Originally published by Springer-Verlag Berlin Heidelberg New York in 1989

2132/3140-543210 – Printed on acid-free paper

"Wir haben gefunden, dass der Grund stehender Gewässer keineswegs allgemein ein Reich der Aufschüttung, sondern zum grossen Teil das Gegenteil davon ist, nämlich ein Reich der Exesion und Denudation" (Arnold Heim 1924, pp. 45-46). *)

Preface

During the so-called Mid-Cretaceous interval, approximately 100 million years ago, the earth experienced a dynamic phase in its geologic history. Enhanced global tectonic activity resulted in a major rearrangment of the continental plates; accelerated spreading rates induced a first-order sea level highstand; intense off-ridge volcanism contributed to a modeled high atmospheric CO_2 rate; climatic conditions fluctuated; and major changes occurred in biologic evolutionary patterns. With the initiation of a gradual change from an equatorial, east-west directed current-circulation pattern to a regime, dominated by south-north and north-south directed current systems, the earth's internal clock was set for Cenozoic, "modern" times.

The Mid-Cretaceous dynamic phase is recorded in a suite of sediments of remarkable similarity around the globe. Shallow-water carbonate platforms drowned on a global scale; widespread sediment-starved, glauconite and phosphate-rich sequences developed; and consequently, pelagic sedimentary regimes "invaded" shelf and epicontinental sea areas.

This typical "deepening-upward" pattern is well-documented in Mid-Cretaceous sequences along the northern Tethys margin. Shallow-water carbonates are overlain by condensed glauconitic and phosphatic sediments, which, in turn, are blanketed by pelagic carbonates.

In this volume, the example of the western Austrian helvetic Alps, built up of inner and outer shelf sediments deposited along the northern Tethys margin, is used to elucidate the paleoceanographic conditions, under which the Mid-Cretaceous triad of platform carbonates, condensed phosphatic and glauconitic sediments, and pelagic carbonates was formed. In the first part, the evolution of this sequence is traced from the demise of the platform (Aptian) to the return of detritus-dominated deposition (Upper Santonian). The second part includes a discussion of the reconstructed paleoceanographic and tectonic variables, their possible interaction, as well as their influence on sediment properties during this period. Special attention is paid to (1) subsidence behavior of the inner, platform-based shelf and the outer shelf beyond the platform, (2) ammonoid paleobiogeography, (3) the northern tethyan current system and its impact on sediment patterns, (4) the influence of an oxygen mi-

*) *We have found that the sea bottom by no means just represents a site of deposition; no, in many cases it represents the opposite, i.e., a site of exesion and denudation* (= weathering and uncovering).

nimum zone, (5) sediment bypassing mechanisms on the inner shelf, (6) condensation processes, (7) phosphogenesis, (8) relative sea level changes, (9) genesis and the development of unconformities, (10) tectonic phases and their impact on sediment configuration, (11) drowning of the shallow-water carbonate platform, and (12) "asymmetric" sedimentary cycles.

The detailed reconstruction of the development of sedimentary patterns both in time and space in this particular area, and its environmental interpretation, given in this volume, may serve as a contribution to a better understanding of the Mid-Cretaceous dynamic phase in earth's history.

This volume is an updated, English version of the stratigraphic-sedimentological part of my Ph.D. thesis, which I completed at the ETH Zürich. I acknowledge with many thanks the support of my advisors Rudolf Trümpy and Hans Rieber (both Zürich), as well as of Rudolf Oberhauser (Vienna), Pieter Ouwehand (Solothurn), Beat Keller (Bern) and Michel Delamette (Fribourg). I am grateful to Robert Garrison (Santa Cruz), who reviewed several drafts of this work, and who provided the necessary infrastructure to complete it. I thank Gerald Friedman and David Kopaska-Merkel (both Troy), and Rick Behl and Kurt Grimm (both Santa Cruz) for their constructive comments. I am indebted to the Swiss National Science Foundation and the ETH Zürich for financial support of my postdoctoral tenure at the University of California, Santa Cruz.

Dr. W. Engel and Springer Verlag are thanked for publication in the Lecture Notes Series.

This work is considered a contribution to the International Geological Correlation Programs 58 ("Mid-Cretaceous Events"), 156 ("Phosphorites") and 198 ("Northern Tethys Margin"), as well as to the evolving Global Sedimentary Correlation Program on Cretaceous Resources, Events, and Rhythms.

Contents

1 Introduction

1.1 Results of this Work and their Relation to the Observations of Arnold Heim

In the first half of this century, the Swiss and cosmopolitan geologist Arnold Heim (1882-1965) repeatedly emphasized the dynamics of sedimentary environments. In an early paper on gravity flows ("subaqueous solifluction"; Heim 1908; also 1924, 1946), Heim anticipated the concept of fluxo- or megaturbidites (cf. Kuenen 1958; Bouma 1987); in his classical paper of 1924, he stressed the importance of bottom-hugging currents in processes of erosion, winnowing, and condensation, and related sedimentary discontinuities to physical and chemical changes in paleoceanographic conditions, avoiding the hitherto common equating of stratigraphic discontinuities with emersion and subsequent transgression (cf. also Heim 1934, 1958; Heim in Baumberger et al. 1907; Heim and Seitz 1934).

Heim made many of his key observations in the course of detailed fieldwork in the eastern Swiss and western Austrian Alps, especially in a sequence of Mid-Cretaceous sediments, which is characterized by a profound deepening-upward trend: Barremian to Lower Aptian shallow-water platform carbonates, overlain by Aptian to Lower Cenomanian condensed glauconitic and phosphatic sediments, overlain by Upper Albian to Upper Santonian pelagic carbonates (Heim 1905, 1909, 1910-1917, 1919, 1923, 1924, 1934, 1958; Heim and Baumberger 1933; Heim and Seitz 1934).

A detailed reexamination of this sequence in the key area of Vorarlberg (western Austria), in progress for the last 10 years (Oberhauser 1982; Bollinger 1986; Föllmi 1981, 1986, 1989; Föllmi and Ouwehand 1987), confirms and supports the approach of Heim as realized more than 60 years ago. Some of the major inferences in the work presented here are:

1. Drowning of the Barremian to Lower Aptian shallow-water carbonate platform was directly related to a landward shift of a bottom-hugging, geostrophic current, importing nutrient-rich waters.
2. Reworking and redeposition of sediments were widespread phenomena, which exerted a large impact on the overall sediment configuration. Especially four short episodes (at the Aptian-Albian, Cenomanian-Turonian, Turonian-Coniacian and Santonian-Campanian boundaries) were dominated by reorganization processes of preexisting sediments. These episodes correlate in part to eo-alpine tectonic events.
3. Unconformities and unconformity-related condensed beds of Aptian to early Cenomanian age are the outcome of a complex, multiple event history, in which current activity played an important role. This genuine type of unconformity should not be confused with single-event "unconformities", which appear at the base of event beds, and may include an erosional hiatus as well.
4. The formation of phosphatic particles, diaclasts, and horizons, and subsequently the concentration of phosphates into condensed beds is interpreted

as the result of the presence of an oxygen minimum zone, of a bottom-hugging current system, and of bypassing sandbeds, which episodically co-vered large biotic communities in a catastrophic way, amongst other, more specific processes.
5. On the verge of the eo-alpine deformation phase, sediment patterns drasti-cally changed along the northern Tethys margin during the Mid-Cretaceous. One of the more prominent changes is given with the termination of cycli-cally influenced sedimentation, which is documented in a suite of Lower Cretaceous, asymmetric, "regression-transgression" sequences.

1.2 Events and Episodes in Mid-Cretaceous Times

In Mid-Cretaceous times, the "French revolution" of the Phanerozoic took place. In the Aptian-Coniacian interval, considered as probably the most eventful period of the Cretaceous, earth's course was set for "modern", Ceno-zoic times (e.g., Reyment and Bengston 1986). Without doubt, tens of millions of years (my) were still to go, until an externally derived cataclysm directly initiated the Cenozoic era, but the fundament to this period was already laid at that time. The C/T-boundary event catalyzed the in the Mid-Cretaceous initiated development by acting as an (almost?) instantaneous *coupe de grâce* to many "doomed", i.e., genetically inert biotic groups, and by creating a large number of ecologic niches.

The "revolution" was driven by a phase of intensified tectonism, inducing increased spreading rates along major ridge systems and widespread off-ridge volcanism (Schwan 1980; Schlanger et al. 1981; Fischer 1984; Savostin et al. 1986; Ziegler 1988). Directly or indirectly related to this phase of tectonism, important climatic, paleoceanographic, and biological changes occurred.

The Atlantic started to open along its northern south part (separation of northwestern Africa and South America) and its northern part. The resulting counterclockwise rotation of the African plate relative to Europe culminated in a continent-continent collision and caused the gradual closure of the Tethys ocean (e.g., Savostin et al. 1986; Reyment and Bengston 1986; Ziegler 1987, 1988; Le Pichon et al. 1988). The opening of the Atlantic passage and the gra-dual closure of the circum-equatorial seaway marked the onset of a gradual switch from east-west current-circulation patterns, dominating the Mesozoic and early Tertiary, to "modern" north-south and south-north directed currents (e.g., Berggren 1982).

Due to intense volcanic activity, CO_2-outgassing rates increased (Berner et al. 1983; Berner and Lasaga 1989; Arthur et al. 1985; cf. also Volk 1987). Along complicated and still not well-understood paths, the increased availabi-lity of CO_2 may have contributed to higher levels of atmospheric CO_2, al-though a portion of it may have been buffered, due to increased rates of C_{org} burial. Simultaneously, the CCD (Calcite Compensation Depth) rose rapidly from about 4 km to a level possibly as shallow as 2 km, affecting the distri-bution of calcium carbonates (in late Aptian; Thierstein 1979; Tucholke and Vogt 1979; Arthur et al. 1985).

Probably related to the increased spreading rates and the resulting increase in ridge volume, the second Phanerozoic first-order, eustatic, sea level cycle approached its maximum during Mid-Cretaceous times (zenith in Turonian; Vail et al. 1977; Fischer 1984; Haq et al. 1987). The resulting large epicontinental seas were probably sources of warm, saline waters, which contributed to salinity stratified oceanic waters and possibly to slow oceanic turnover rates, resulting in lower oxygen contents within bottom waters and episodic anoxia (Arthur and Jenkyns 1981; Brass et al. 1982; Bralower and Thierstein 1984).

The Mid-Cretaceous witnessed important biological changes, such as a first-order increase in diversity of planktonic calcareous foraminifera (e.g., Caron 1985), and the rapid evolution of angiosperms, replacing gymnosperms, ferns, and their allies (e.g., Hickey 1984; Tschudy 1984). Two Mid-Cretaceous episodes of rapid evolutionary changes in the group of marine biota such as benthic foraminifera, scleractinian corals and ammonoids are included in the periodic suite of major extinction "events" of Raup and Sepkoski (1986, at the Aptian-Albian and Cenomanian-Turonian boundaries; compare also Kauffman 1984; Sepkoski 1989).

The Mid-Cretaceous was also a time of (still poorly defined) climatic variations. The Barremian and early to early late Aptian periods were presumably dominated by high average temperatures (e.g., Mutterlose 1987), whereas the late Aptian and early Albian period experienced a stronger cooling phase (Juignet et al. 1973; Vakhrameev 1978; Alvarez-Ramis et al. 1981; Bollinger 1986; Ouwehand 1987). Kemper (1983, 1987) even claimed the presence of ice caps during this episode (cf. Frakes and Francis 1988; Rich et al. 1988). Subsequently, temperatures increased, and the middle and late Albian is generally considered as a warm period (e.g., Berggren 1982; Schneider et al. 1985). In the latest Albian and Cenomanian, a second, not well-defined cooling period may have occurred (Berggren 1982; Kauffman 1984; Spicer 1987; Crowley and North 1988). In the Turonian and/or Coniacian, average temperatures rose again or remained stable (Berggren 1982; Crowley and North 1988). This interval was probably punctuated by short cooling episodes (Kauffman 1984). In the Santonian or Campanian, a gradual but distinct cooling phase began that persisted into the Paleocene (e.g., Berggren 1982; Spicer 1987).

The above described global paleoceanographic, climatic and biological changes during the Mid-Cretaceous are recorded in shelf sediments in a surprising consistency throughout the world.

Shallow-water platform carbonates of Barremian to Albian age are widespread throughout tropical and temperate areas (e.g., Pacific and Atlantic, Gulf of Mexico, Western Interior, Florida, along the northern Tethys margin; cf. compilation in Schlager 1981). During Aptian and Albian, the vast majority of these carbonate platforms drowned. This encompassing episode has been referred to by Schlager (1981; pp. 197, 205) as "global mass extinctions of reefs and platforms".

Aptian to Cenomanian, detritus-rich sediments, associated with glauconitic and phosphatic sediments are equally widespread, and occur along the northern Tethys margin from Spain to the western Carpathians and the southern

Soviet Republics, along the southern Tethys margin from the Himalayas to Tunisia and Marocco, in boreal areas such as England, northwestern France, northwestern Germany, Canada, and in southern temperate areas such as South Africa and Madagascar (Ouwehand 1987). This "facies" type has commonly been coined with the term *Gault*, a term originally used by English quarryworkers and miners of the 18th century, and subsequently (19th century) transferred to occurrences throughout the world.

The widespread distribution of Cenomanian and Turonian pelagic sediments in marginal and epicontinental settings is known for over a century ("Cenoman-Transgression" of E. Suess 1883).

An "ideal" or at least typical Mid-Cretaceous sequence, therefore, begins with shallow-water carbonates, followed by detritus-rich, more or less condensed sediments, commonly incorporating authigenic suites of minerals (e.g., glauconite, francolite), and ends with pelagic sediments.

1.3 Documents of the Mid-Cretaceous in Western Austria

A well-developed example of the Mid-Cretaceous triad of shallow-water platform carbonates, condensed phosphatic and glauconitic sediments, and pelagic carbonates is present along the southern European, northern Tethys margin, where it extends from southeastern Spain to the western Carpathians (e.g., Delamette 1988a). As part of this zone, the western Austrian helvetic Alps are especially intriguing, because they include a Mid-Cretaceous transect <u>across</u> a carbonate platform margin (Figs. 1 and 2). A similar and correlatable situation is given in southeastern France (Vocontian Trough; e.g., Delamette 1988b).

Mid-Cretaceous sediments of the western Austrian helvetic Alps include:

1. Barremian to Lower Aptian shallow-water platform carbonates (**Schrattenkalk Formation**), constituting an inner shelf segment and an approximately coeval sequence of alternating carbonates and marls (**Drusberg Formation**), deposited beyond the platform margin (designated here as outer shelf).
2. Lower Aptian to Upper Albian or Lower Cenomanian, condensed glauconitic and phosphatic sandstones, mud and marlstones (platform-based; **Garschella Formation**); channel and fan systems, consisting of redeposited, platform-derived sediments (along a steepened platform margin; Garschella Formation) and hemipelagic mud and marlstones (outer shelf; **Mittagspitz Formation** and Garschella Formation; Fig. 2).
3. Uppermost Albian to Upper Santonian pelagic carbonates, including gravity flow deposits of reworked, older sediments (**Seewen Formation**; Fig. 2). The Seewen Formation is overlain by Upper Santonian to Upper Campanian detritus-rich marls and mudstones (**Amden Formation**), which are briefly discussed in this volume.

The platform-carbonate sediments of the Schrattenkalk Formation represent the youngest member in a suite of Lower Cretaceous shallow-water carbonate platforms, which evolved and demised in a cyclic pattern (Fig. 3; Sect. 3.9).

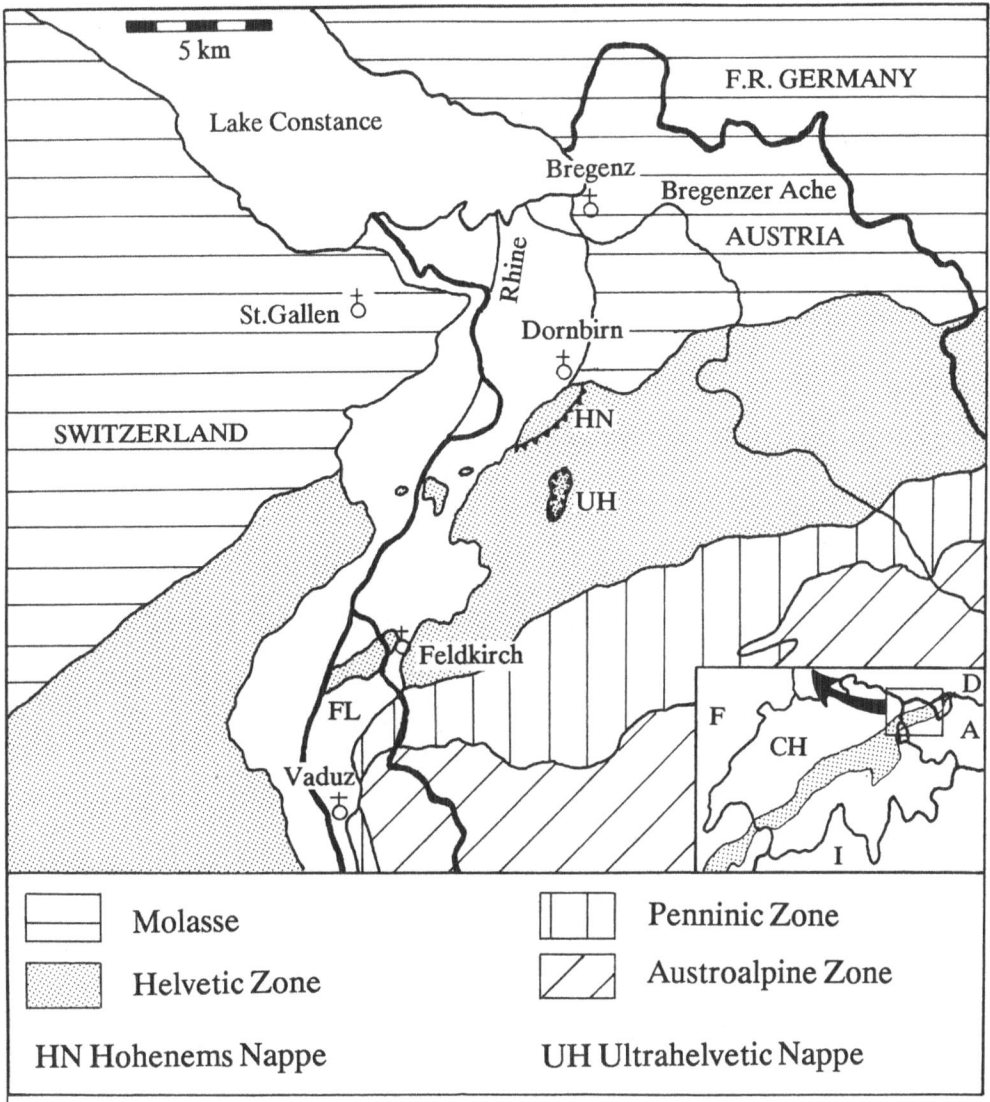

Fig. 1. Tectonic map of the eastern Swiss and Vorarlberg helvetic area (after Spicher 1980). The main tectonic unit in the Vorarlberg helvetic Alps is represented by the Säntis Nappe. Along the northwestern rim, a deeper, more proximal nappe crops out (Hohenems Nappe). On top of the Säntis Nappe, small ultrahelvetic nappe remnants are preserved, the most important being located in the area of the Hohe Kugel (indicated by UH; after Oberhauser 1982; Wyssling 1985)

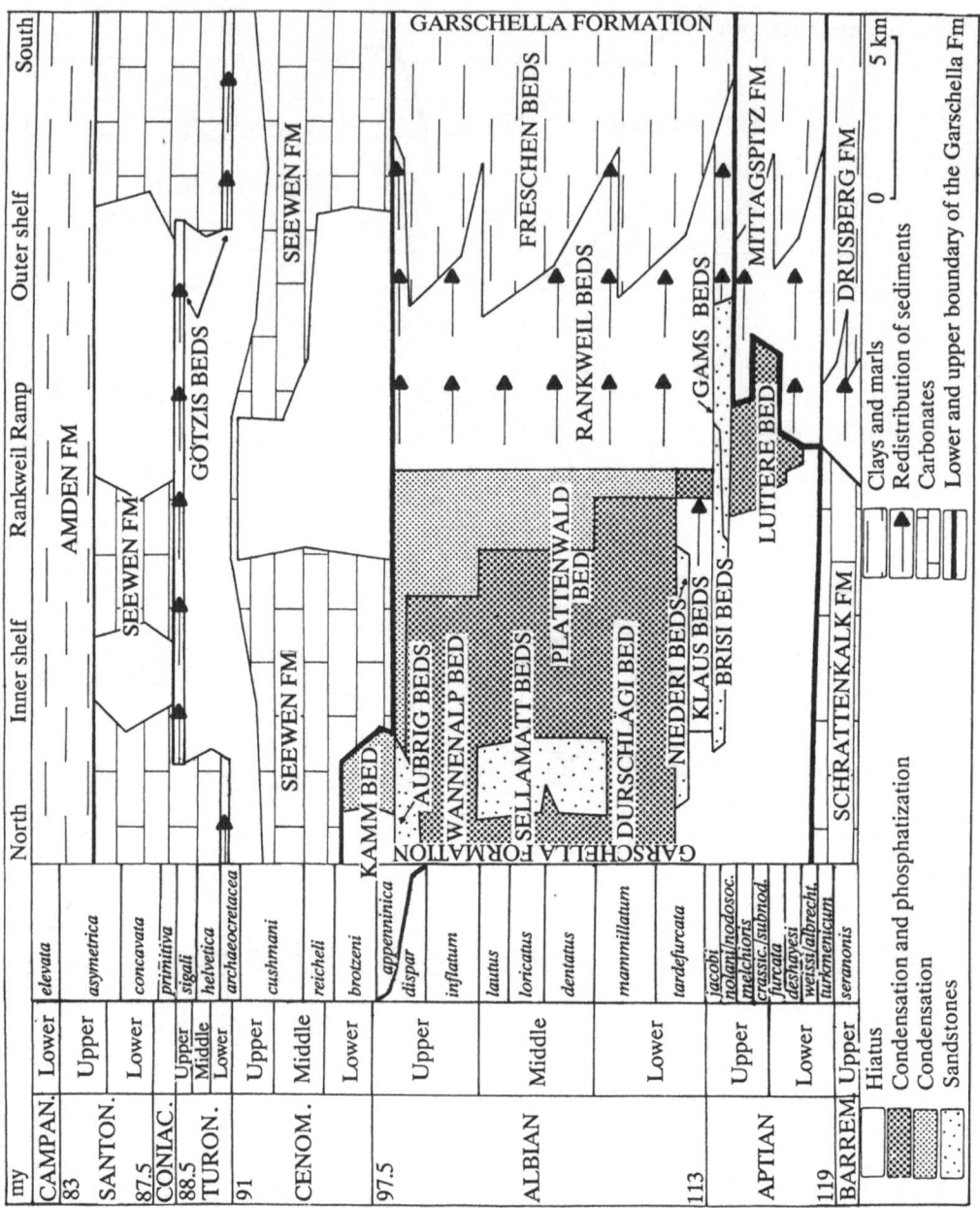

Fig. 2. Time-space distribution of Barremian-Campanian sediments in the Vorarlberg helvetic Säntis Nappe. The biozonation used in the figure and in this volume is based on ammonoids (*seranonis* to *dispar* Zone; Mikhailova 1979; Druschtchitz and Gorbatschik 1979; Casey 1961; Owen 1971, 1975) and planktonic foraminifera (*appenninica* to *elevata* Zone; Robaszynski and Caron 1979; Caron 1985). Timetable is adopted from Harland et al. (1982)

In the Aptian, major paleoceanographic and tectonic changes prevented the further development of shallow-water carbonates, and induced a current-dominated sedimentary environment on the inner shelf. A complex system of condensed phosphatic and glauconitic sediments resulted, preserved in the Garschella Formation. In the latest Albian and early Cenomanian, a pelagic regime gradually emerged, and the heterogeneous sediments of the Garschella Formation became blanketed with calcareous oozes (Seewen Formation). This regime reached its zenith in the Turonian. Subsequent decline led to a time-transgressive change into a detritus-dominated sedimentary environment (Amden Formation).

The evolution of the triad of Schrattenkalk, Garschella, and Seewen Formations was punctuated by six episodes of retarded sedimentation, erosion, and stratal condensation, and/or major erosion and redeposition. Two episodes occurred in the period of early to early late Aptian, and in the latest Aptian and contributed to the demise of the Schrattenkalk carbonate platform (Sect. 3.8). A major episode of ultralow sediment accumulation rates, stratal condensation and reworking took place during the Albian, and is recorded in an "ultra"-condensed phosphatic bed (net sediment accumulation rates of 2-20 cm/my over at least 10 my; Sects. 2.6 and 3.6). Further episodes of major erosion and redistribution of eroded sediments are recognized in the latest Cenomanian to earliest Turonian, late Turonian to Coniacian, and late Santonian to early Campanian time intervals.

1.4 Data Sources

Inferences made in this work are based on stratigraphic, sedimentologic, and paleontologic data from approximately 300 different measured sections and examined localities in Vorarlberg, western Austria. The sections are documented in detail in Föllmi (1986).

A precise biostratigraphic framework is provided by abundant ammonoids and inoceramid bivalves, collected in strata of the Garschella Formation, as well as by globotruncanid foraminifera, observed in the Seewen and Amden Formations (Fig. 2; Föllmi 1986, 1989; Föllmi and Ouwehand 1987).

For general information on the Mid-Cretaceous helvetic unit, the reader is referred to Delamette (1985, 1988a,b; southeastern France and western Switzerland), Korner (1978; central Switzerland), Ouwehand (1987; eastern Switzerland), and Gebhard (1983, 1985; Vorarlberg, Allgäu), as well as to the valuable work of Arnold Heim (1905, 1909, 1910-1917, 1919, 1923, 1924, 1934, 1946, 1958; Heim and Baumberger 1933; Heim and Seitz 1934), Ganz (1912), Fichter (1934), Schaub (1936, 1948), and Bolli (1944).

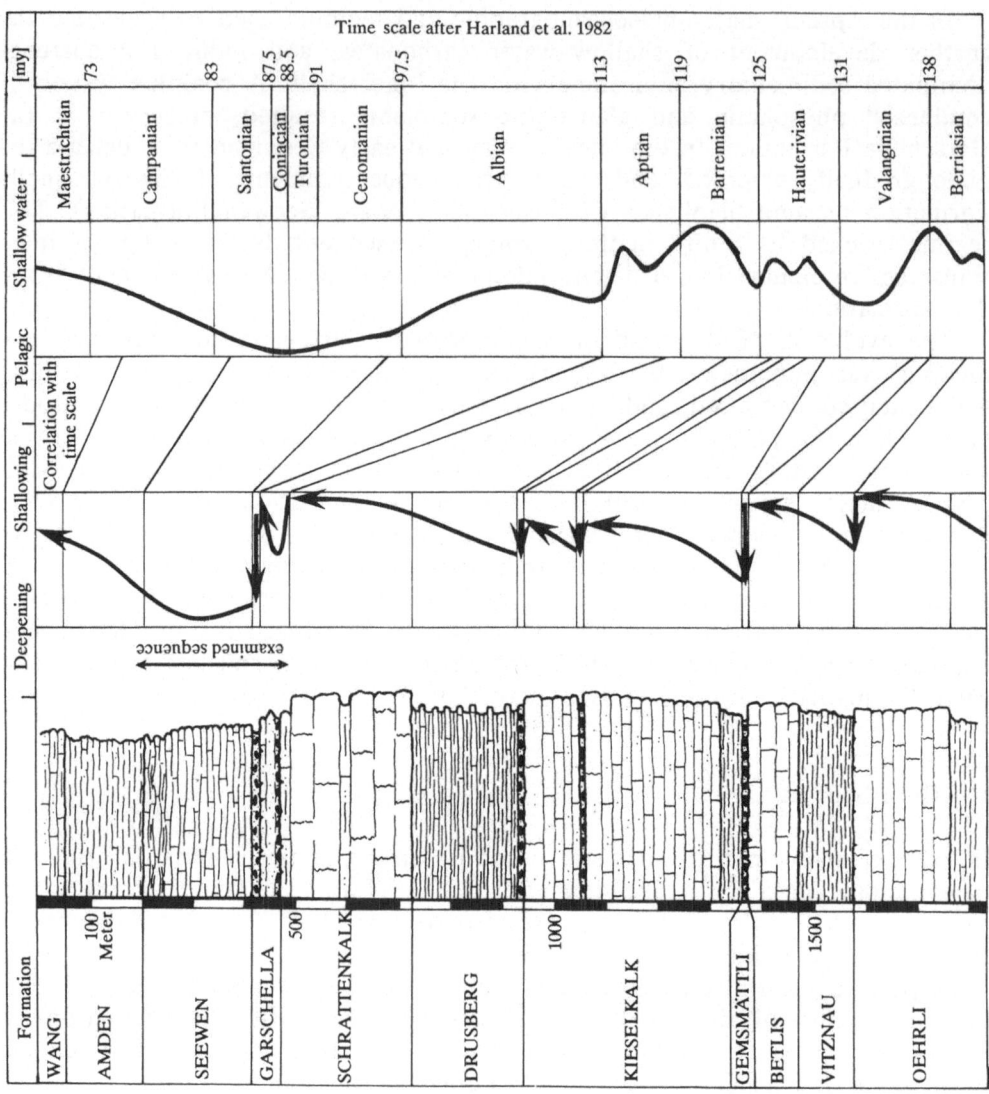

Fig. 3. Generalized Cretaceous section of the eastern Swiss helvetic zone (after Funk 1971; Trümpy 1980; Burger 1985; Wyssling 1986). The shallowing-upward sequences include generally a terrigenous lower and a calcareous upper part. The deepening-upward sequences consist of thin and condensed, commonly phosphatic beds. The right column in the figure shows a tentative, relative sea level curve, based on facies changes

2 Distribution and Interpretation of Mid- and Lower Upper Cretaceous Helvetic Shelf Sediments in Vorarlberg (Western Austria)

2.1 Introduction

In this section, a time-based overview is given over the evolution of sediment configurations, as they are preserved in the helvetic Alps of western Austria. The time intervals chosen here (late Barremian to early Aptian, middle early to early late Aptian, middle late Aptian, latest Aptian to earliest Albian, early to late Albian, latest Albian to earliest Cenomanian, middle to late Cenomanian, latest Cenomanian to early Turonian, middle Turonian, late Turonian to Coniacian, late Coniacian to late Santonian, late Santonian to early Campanian) are considered to represent semi self-contained intervals or phases in the evolution of the helvetic shelf.

2.2 Late Barremian to Early Aptian

In Barremian time, an ultimate carbonate platform evolved within the early Cretaceous cycle of pro- and retrograding shallow-water carbonate platforms (Schrattenkalk Formation; Figs. 2 and 3; Sect. 3.9.1). The Schrattenkalk platform attained its maximum extent in late Barremian or earliest Aptian times (Bollinger 1986; Fig. 56). Platform sediments from this episode are traceable along the entire northern Alps, and extend into southeastern France, Spain, and the Carpathian mountains (Lienert 1965; Zacher 1973; Funk and Briegel 1979; Arnaud-Vanneau 1980; Bollinger 1986). Eastward telescoping of facies zones within the Schrattenkalk and in the overlying Garschella Formations suggests a narrowing and steepening of the platform toward the eastern part of the helvetic shelf (Sect. 3.2.1; Zacher 1973; Föllmi 1986).

Consisting predominantly of calcarenitic bio- and oopelsparites, and micrites with abundant remains of miliolids, orbitolinids, and dasycladacea, this carbonate platform facies reflects deposition in shallow water. Stromatoporoid-rudist-scleractinian coral patch reefs developed in inner platform areas, whereas ooid shoals formed in more distal settings (detailed descriptions in Scholz 1979, 1984; Bollinger 1986; cf. also Zacher 1973; Arnaud-Vanneau 1980).

Commonly, the platform sediments have gone through cycles of deposition and remobilization (initiated most probably by storm events). Carbonate particles were transported to distal areas as far as the muddy outer shelf, where they form calcarenitic banks within the Drusberg Formation. The regular distribution of redeposited sediments in the proximal outer shelf and the absence of larger erosive structures (e.g., channels) suggest a gentle, homoclinal ramp that connected the platform with the outer shelf (Figs. 2, 3, and 49; Heim and Baumberger 1933; Zacher 1973; Felber and Wyssling 1979; Bollinger 1986).

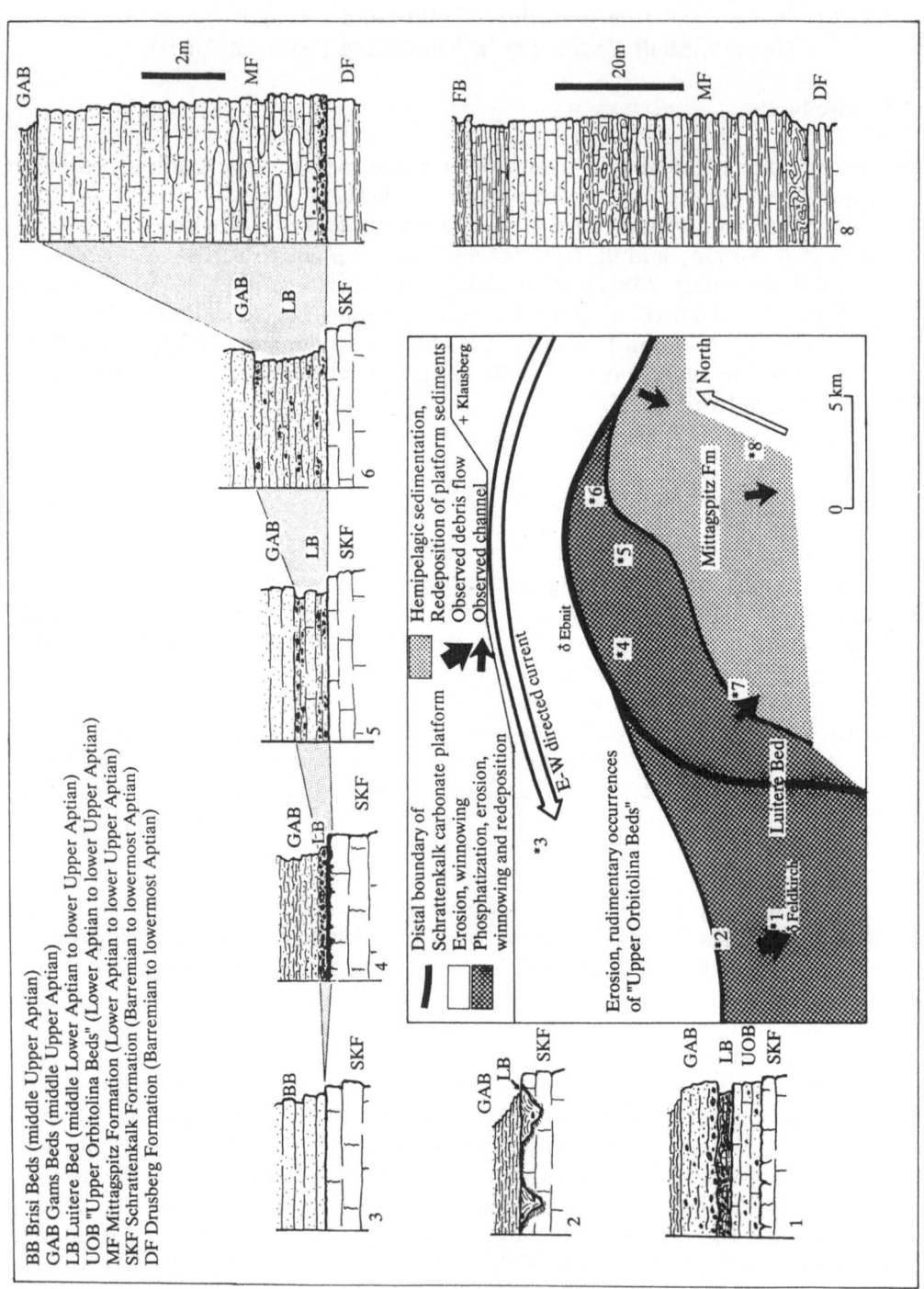

BB Brisi Beds (middle Upper Aptian)
GAB Gams Beds (middle Upper Aptian)
LB Luitere Bed (middle Lower Aptian to lower Upper Aptian)
UOB "Upper Orbitolina Beds" (Lower Aptian to lower Upper Aptian)
MF Mittagspitz Formation (Lower Aptian to lower Upper Aptian)
SKF Schrattenkalk Formation (Barremian to lowermost Aptian)
DF Drusberg Formation (Barremian to lowermost Aptian)

2.3 Middle Early to Early Late Aptian

2.3.1 General Overview

The middle early to early late Aptian time interval is recorded either by a hiatus in the sediment column, or by three different, yet approximately time-equivalent types of sediment (Figs. 2, 4, and 6).

1. Proximal areas of the Schattenkalk platform, exposed in central and eastern Switzerland, were blanketed by sandy, crinoid-rich marls and muds (Lower to lower Upper Aptian **"Upper Orbitolina Beds"**; max. 25 m; Lienert 1965; Keller 1983; Bollinger 1986). The distal platform part, preserved in Vorarlberg, generally lacks "Upper Orbitolina Beds" (exception: section 1 in Fig. 4). Instead, a distinct, unconformable, and erosive boundary between the Schrattenkalk Formation and younger Garschella Formation sediments indicates a hiatus, suggestive of persistent conditions of zero net sediment accumulation rates, as well as of erosion of the uppermost Schrattenkalk Formation sediments (see below; Figs. 2, 4, and 6; section 3 in Fig. 4).
2. Along the platform margin and within the proximal outer shelf settings beyond the carbonate platform, heterogeneous, condensed, phosphatic sediments formed (middle Lower to lower Upper Aptian **Luitere Bed**; max. 1.5 m; Figs. 2, 4, and 6). The Luitere Bed phosphatic sediments display four distinct facies types:

   A. The most common facies type starts with a thin (<5 cm), autochthonous phosphatic crust on top of the Schrattenkalk or Drusberg Formation, which commonly penetrates the calcareous substrate along fissures. The crust consists of intimately interbedded phosphatized *Hedbergella* micrites and phosphatized microbial mats (Krajewski 1984; Föllmi 1986; Fig. 5A).
   The phosphatic crust is covered by a thin (<30 cm), multi-event winnowed, condensed, nodular, phosphatic bed (Föllmi et al. in press). The

Fig. 4. Middle early to early late Aptian.
Palinspastic map of the Vorarlberg Säntis nappe displays the distribution of the Lower to lower Upper Aptian "Upper Orbitolina Beds", Luitere Bed, and Mittagspitz Formation. Note the bight-shaped margin of the drowning Schrattenkalk platform, the limitation of the Luitere Bed sediments to the proximal outer shelf area (beyond the drowning platform) in the east and the onlap of these sediments onto the distal platform area in the west. Due to differential subsidence, the Luitere Bed-covered marginal platform part was transformed into a proximal part of the outer shelf zone during early and early late Aptian times (Figs. 7, 16, and 19; Sect. 3.2). Correlation of Lower to lower Upper Aptian sediments is indicated by dots. Note different scale for section 8 (2-m scale bar for sections 1-7; 20-m scale bar for section 8)

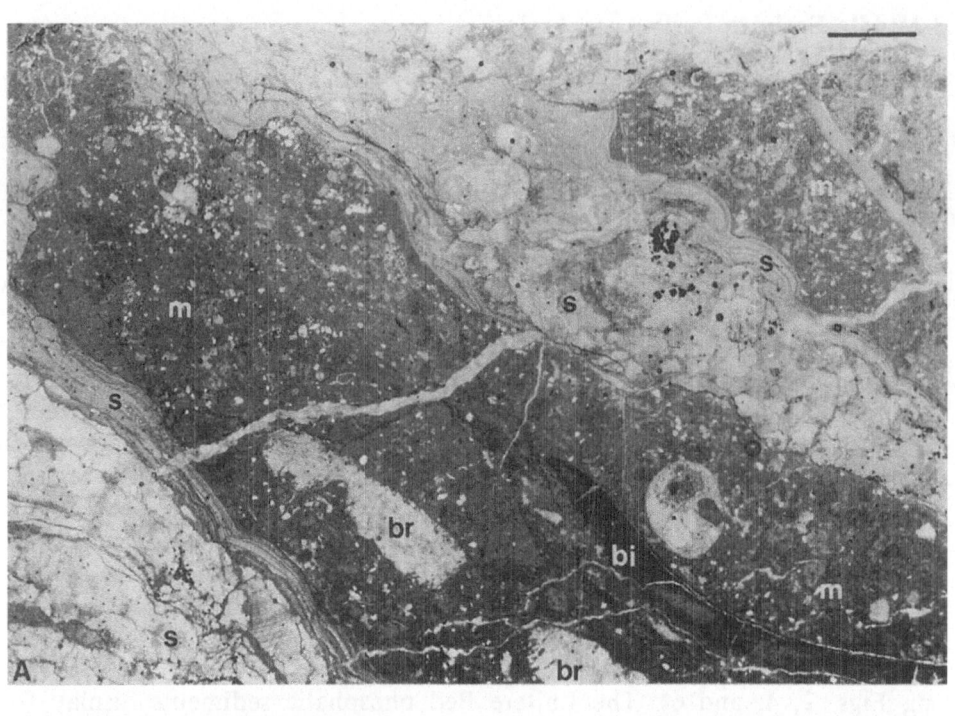

phosphatic particles consist of phosphatized fossil remains (brachiopods, bivalves, gastropods, ammonoids, belemnites, echinoids), as well as of phosphatized lithoclasts (e.g., rip-up clasts from the subjacent Schrattenkalk Formation). The phosphatized particles display several distinct phosphate generations (Sect. 3.4), and are embedded in glauconitic sandstones and marls, identical to superposed Gams Beds (see below). This facies type is typical and commonly encountered in the Luitere Bed localities (section 4 in Fig. 4).

B. A second facies type, limited to the area of Feldkirch, is represented by an allochthonous, inversely to normally graded, nodular, phosphatic bed. The phosphatic "diaclasts" (Einsele et al. in press) consist of well-preserved fossils (e.g., sponges, brachiopods, bivalves, gastropods, ammonoids, belemnites, echinoids). Included Schrattenkalk pebbles and boulders commonly display a phosphatized surface and occasionally a phosphatized rim of *Hedbergella* micrites (section 1 in Fig. 4).

The local presence of this facies type on top of a thin sequence of "Upper Orbitolina Beds" ensures its allochthonous character. The deposition is interpreted as a debrite (debris flow deposition), probably derived from localities in which Luitere Bed sediments of the above discussed facies type A were present (Fig. 4).

C. A third facies type, observed in the area northwest of Feldkirch, is restricted to pocket and fissure infillings, present on top of the uppermost Schrattenkalk bed. The depressions have been colonized by columnar microbial mats (preserved in a nonphosphatized state; Fig. 5B). Intervening void spaces are infilled with *Hedbergella* micrites, identical to micrites preserved in facies types A and B. Schrattenkalk sediments sub- and adjacent to the depressions display a phosphatized surface (section 2 in Fig. 4).

D. A fourth facies type is limited to areas immediately beyond the Schrattenkalk platform margin and consists of detritus and *Hedbergella*-rich marls (max. 1.5 m). The sediments include abundant, 0.1-0.3 mm-sized, glauconitized and phosphatized particles (peloids, phosphatized fossil fragments, and Schrattenkalk lithoclasts; section 6 in Fig. 4). Transitions to facies type A (section 5 in Fig. 4) as well as to the Mittagspitz Formation (see below; Fig. 6) have been observed.

Fig. 5. Thin-section photomicrographs of Luitere Bed sediments (bar = 1 mm).
A. Pristine phosphatic crust, consisting of phosphatized microbial colonies (s; internally recrystallized) and phosphatized *Hedbergella* micrites (m; with abundant siliciclastic detritus). br = bored brachiopod fragments; bi = phosphatized bivalve fragments (Luitere Bed; facies type A).
B. Depression on top of the Schrattenkalk Formation (sk; biopelmicrite), overgrown by nonphosphatized microbial colonies. Fissures in upper part of microbial colonies are infilled with *Hedbergella* micrites (m). The Schrattenkalk surface is bored (b) and slightly phosphatized (Luitere Bed; facies type C)

3. In the distal prolongation of the Luitere Bed facies D, a sequence of hemipelagic, *Hedbergella*-rich marls and carbonates is present (Lower to lower Upper Aptian **Mittagspitz Formation**; Figs. 2, 4, and 6; sections 7 and 8 in Fig. 4; Felber and Wyssling 1979; Bollinger 1986). The Mittagspitz Formation encompasses less than one to a few meters of sediment in proximal occurrences; in distal localities, the formation includes up to 60 meters of hemipelagic sediments, which are repeatedly interbedded with event strata consisting of redeposited, locally channelized siliciclastic detritus, crinoid calcarenites, phosphatic diaclasts, and eroded Schrattenkalk sediment particles.

2.3.2 Time-Space Relations

Time-space distribution patterns of the Luitere Bed are illustrated by phosphatized ammonoids, which show the following trends in the development of the phosphatic sediments (from east to west, approximately parallel to the shelf edge):

1. Luitere Bed sediments in the easternmost part of the helvetic shelf (Allgäu, southeastern F.R.G.) contain ammonoids of the earliest to early late Aptian period (e.g., *Prodeshayesites*, *Deshayesites*, *Dufrenoyia*: *turkmenicum* to *melchioris* Zone; Heim 1919; Gebhard 1983, 1985; Fig. 2).
2. In Vorarlberg, Luitere Bed ammonoids indicate middle early to early late Aptian (e.g., *Deshayesites*, *Dufrenoyia*, *Cheloniceras*, *Colombiceras*, *Parahoplites*: *deshayesi* to *melchioris* Zone; Heim and Seitz 1934; Föllmi 1986, 1989).
3. In the eastern and central Swiss Alps, sedimentation of Luitere Bed sediments lasted from late early Aptian to early late Aptian (*Cheloniceras*, *Colombiceras*, *Parahoplites*: *furcata* to *melchioris* Zone; Jacob and Tobler 1906; Fichter 1934; Keller 1983; Rick 1985; Ouwehand 1987; Föllmi and Ouwehand 1987).

In Vorarlberg, the Luitere Bed generally covers sediments of the Schrattenkalk and Drusberg Formations (Figs. 2 and 4), whereas the more proximally situated Luitere Bed in Switzerland rests on "Upper Orbitolina Beds" (Fichter 1934; Lienert 1965; Keller 1983).

It follows that the base of the Luitere Bed is diachronous and becomes younger from the east to the west (approximately parallel to the shelf edge), as

Fig. 6. Early to early late Aptian.
Diagrammatic reconstruction perpendicular to the shelf. Note concentration of phosphates along the Schrattenkalk platform margin, onlapping onto the platform. Tentative sediment accumulation profile along the shelf indicates amounts of net sediment accumulation during this period

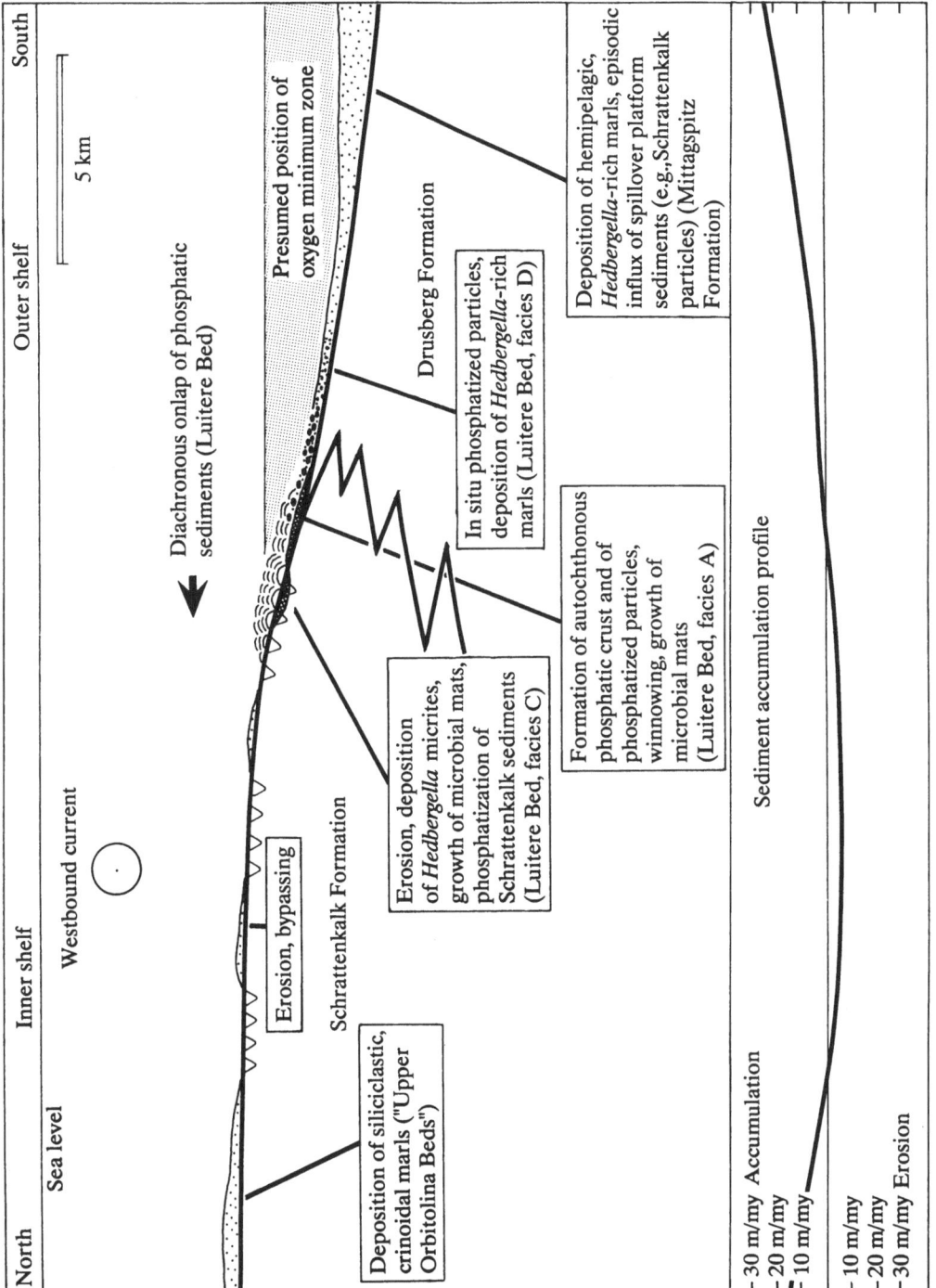

North / Sea level / Inner shelf / Westbound current / Outer shelf / South / 5 km

Diachronous onlap of phosphatic sediments (Luitere Bed)

Presumed position of oxygen minimum zone

Erosion, bypassing

Schrattenkalk Formation

Drusberg Formation

Deposition of siliciclastic, crinoidal marls ("Upper Orbitolina Beds")

Erosion, deposition of *Hedbergella* micrites, growth of microbial mats, phosphatization of Schrattenkalk sediments (Luitere Bed, facies C)

Formation of autochthonous phosphatic crust and of phosphatized particles, winnowing, growth of microbial mats (Luitere Bed, facies A)

In situ phosphatized particles, deposition of *Hedbergella*-rich marls (Luitere Bed, facies D)

Deposition of hemipelagic, *Hedbergella*-rich marls, episodic influx of spillover platform sediments (e.g.,Schrattenkalk particles) (Mittagspitz Formation)

Sediment accumulation profile

30 m/my Accumulation
20 m/my
10 m/my

10 m/my
20 m/my
30 m/my Erosion

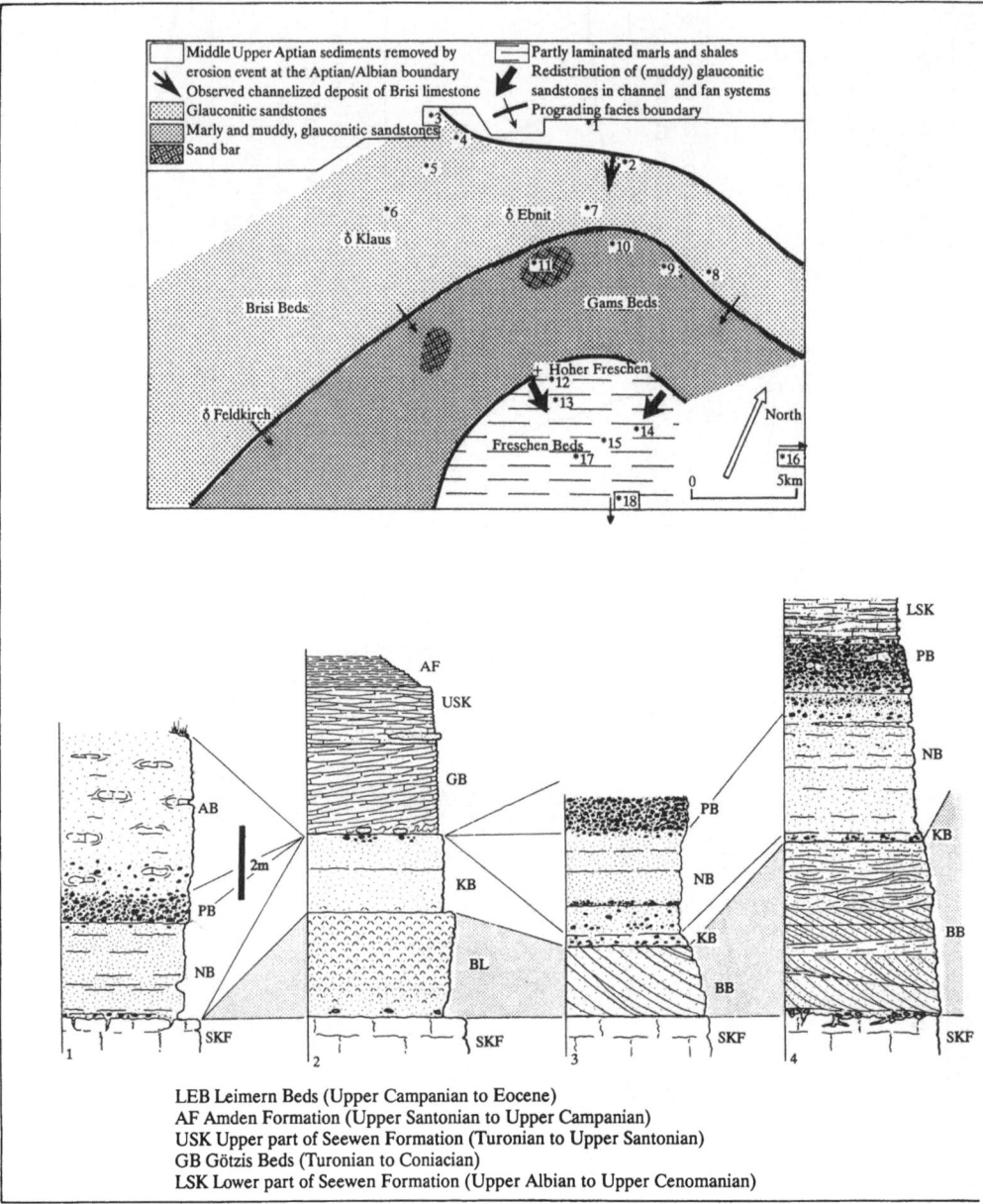

Fig. 7. Middle late Aptian.
Palinspastic map shows the distribution of the Brisi, Gams, and Freschen Beds within the Vorarlberg Säntis Nappe. Absence of middle Upper Aptian sediments in northern areas is due to an erosive phase near the Aptian-Albian boundary. Note position of the only known channelized deposit of Brisi Limestone (section <u>2</u>) and the prograding facies boundary between the Brisi and

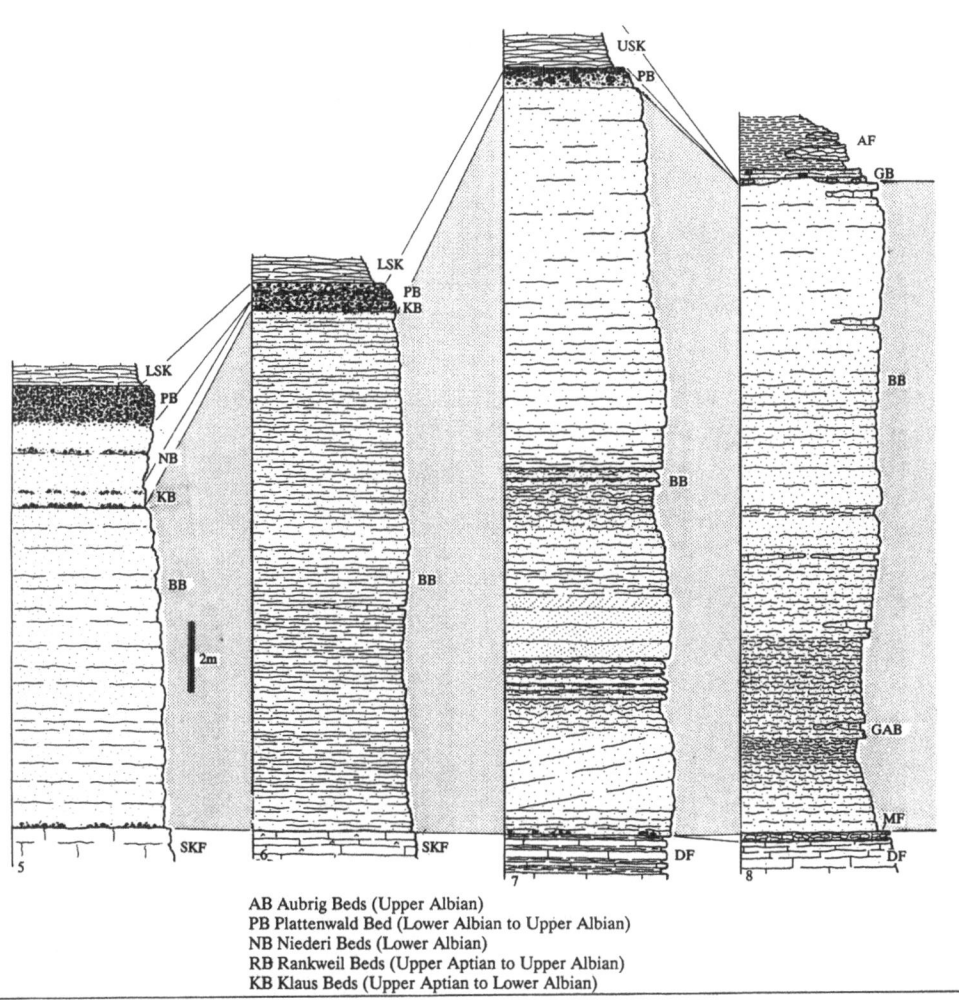

AB Aubrig Beds (Upper Albian)
PB Plattenwald Bed (Lower Albian to Upper Albian)
NB Niederi Beds (Lower Albian)
RB Rankweil Beds (Upper Aptian to Upper Albian)
KB Klaus Beds (Upper Aptian to Lower Albian)

Gams Beds. Within the Gams Beds distribution area, isolated sandbodies are present, connected with the Gams Beds facies by broad transition zones (>100 m; section 11). Morphology and lithology of channel and fan systems within proximal Freschen Beds is shown in Fig. 11. Correlation of middle Upper Aptian sediments is indicated by dots. Note different scale for section 16 (2-m scale bar for sections 1-15, 17 and 18; 5-m scale bar for section 16)

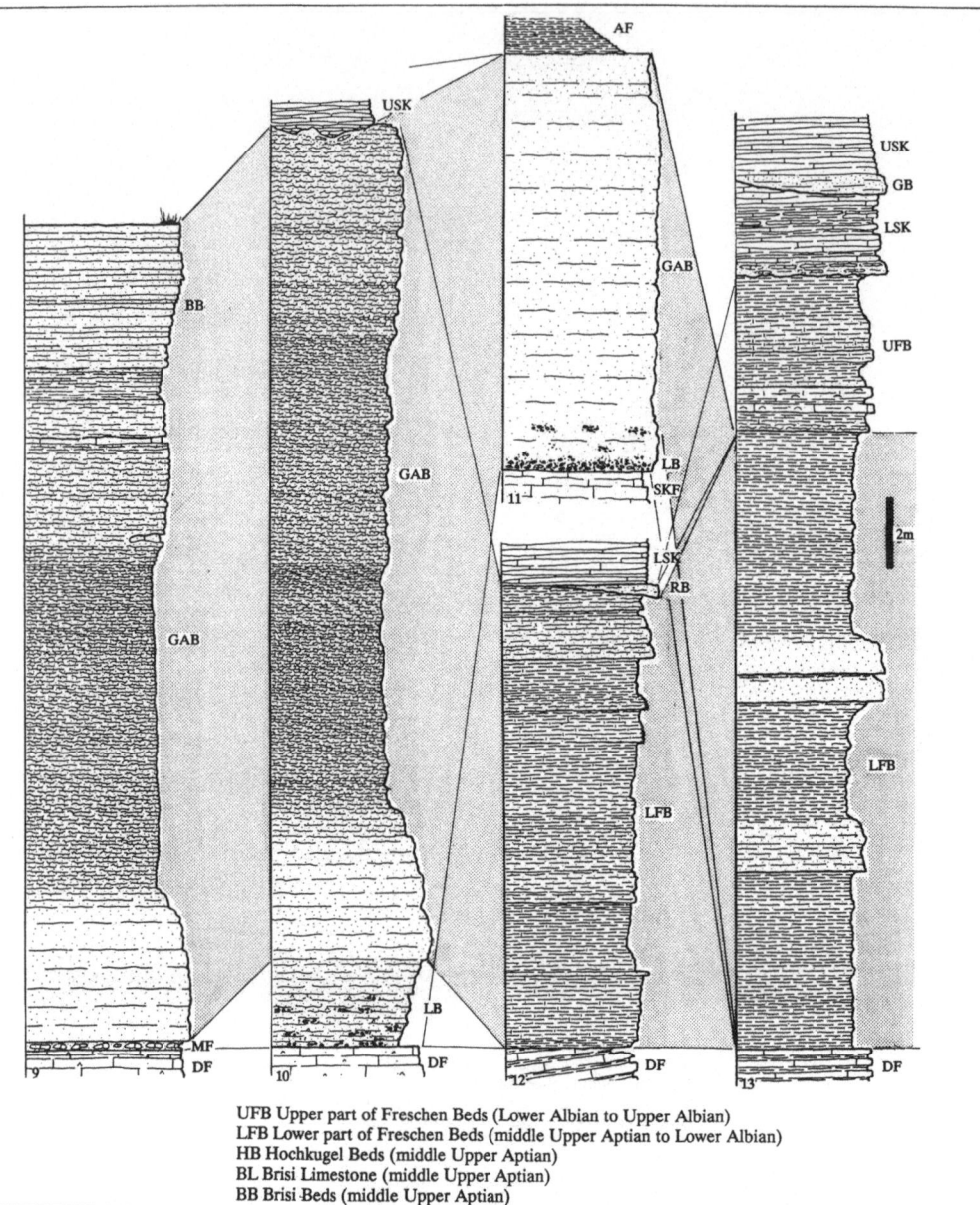

UFB Upper part of Freschen Beds (Lower Albian to Upper Albian)
LFB Lower part of Freschen Beds (middle Upper Aptian to Lower Albian)
HB Hochkugel Beds (middle Upper Aptian)
BL Brisi Limestone (middle Upper Aptian)
BB Brisi Beds (middle Upper Aptian)

Fig. 7. (continued)

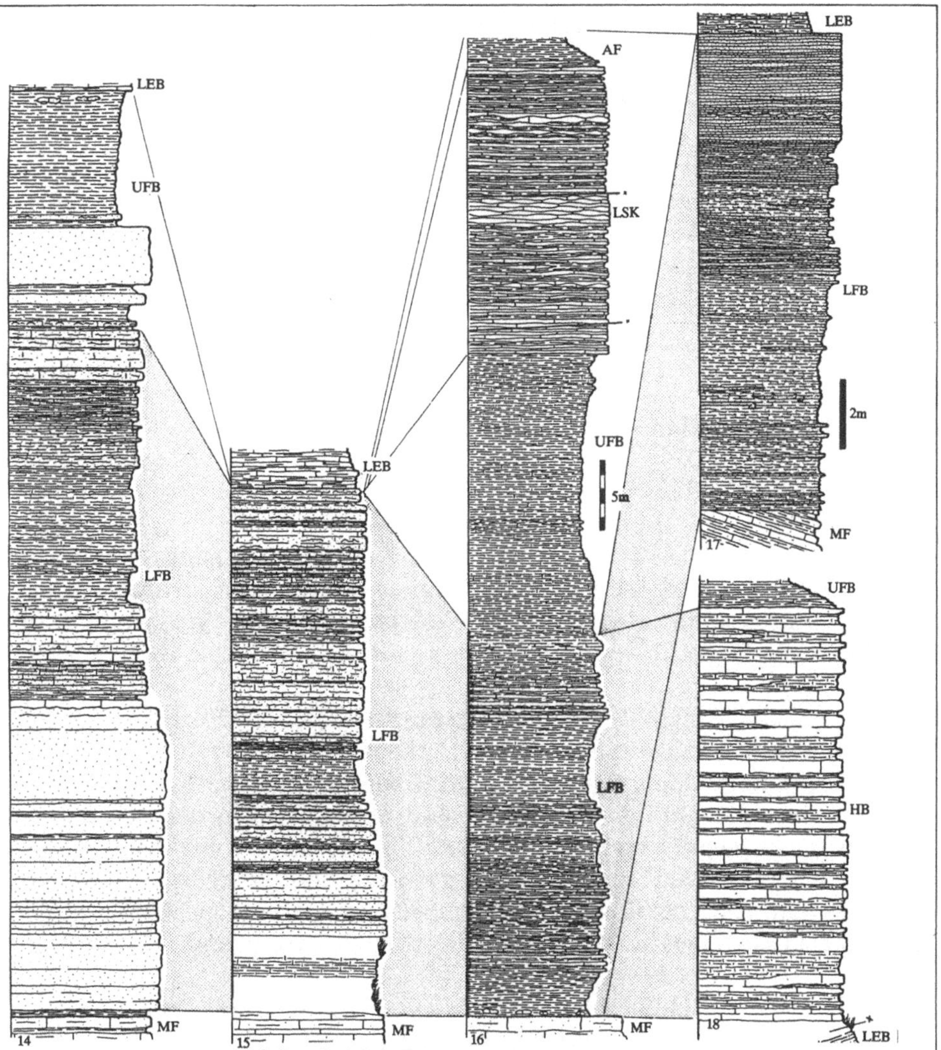

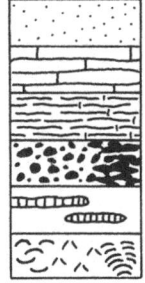

LEB
UFB
LFB
MF

AF
LSK
LFB
UFB
5m
LFB
MF

LEB
LFB
MF

LEB
LFB
2m
MF
UFB
HB
LEB

'14   '15   '16   '17   '18

GAB Gams Beds (middle Upper Aptian)
LB Luitere Bed (middle Lower Aptian to lower Upper Aptian)
MF Mittagspitz Formation (Lower Aptian to lower Upper Aptian)
SKF Schrattenkalk Formation (Barremian to lowermost Aptian)
DF Drusberg Formation (Barremian to lowermost Aptian)

Siliciclastic and glauconitic sands; sandy

Carbonates; calcareous

Clays, marls; muddy

Phosphatic particles, crusts; phosphatic

Chert layers, nodules

Abundant preserved bivalves, echinoderms, microbial mats

well as from distal to more proximal shelf settings, and that the Luitere Bed top approximates an isochron, i.e., the Luitere Bed condensation and phosphatization ended within the same ammonite zone (*melchioris* Zone; Fig. 2).

"Upper Orbitolina Beds" and Mittagspitz Formation sequences are poorly dated and in need of detailed biostratigraphic work. *Hedbergella* (Caron in Bollinger 1986), rare occurrences of ammonoids (*Colombiceras*: at the base of the Mittagspitz Formation in section 7 of Fig. 4; Föllmi 1986, 1989; *Dufrenoyia* in Rick 1985), and age dating in the underlying Schrattenkalk sediments (orbitolinids; Bollinger 1986) and overlying or laterally time-equivalent sediments in the Luitere Bed suggest an early to early late Aptian age for both sequences.

2.4 Middle Late Aptian

2.4.1 General Overview

In the *melchioris* Zone (middle late Aptian; Fig. 2), sediment patterns on the helvetic shelf changed radically. The regime of retarded sediment accumulation, episodic erosion, and phosphogenesis was replaced by a regime of predominantly clastic sediment accumulation (Figs. 4, 6, 7, and 10).

1. In proximal settings of the inner, platform-based shelf, coarse (0.2-2 mm) and commonly tangentially cross-stratified calcarenites accumulated (biosparites; middle Upper Aptian **Brisi Limestone**; max. 30 m; Keller 1983; Ouwehand 1987; Föllmi and Ouwehand 1987). The carbonate clasts consist of fossil fragments (crinoids, bryozoa, brachiopods, bivalves, and algae), abundant micritized Schrattenkalk particles and lithoclasts (reworked), as well as darkened compound particles, similar to black pebbles described by Strasser (1984). Quartz and glauconite particles are present in addition (Fig. 8).

   In Vorarlberg, "autochthonous" Brisi Limestone sediments are almost completely absent. Instead, Lower Albian sediments rest with a distinct unconformity on Schrattenkalk sediments (Fig. 2; section 1 in Fig. 7). An exception is seen in section 2 in Fig. 7, interpreted here as a large channel infill (40-60 m wide) of transported Brisi Limestone sediments, replacing Brisi Beds (see below) and deposited directly upon Schrattenkalk sediments. The presence of large amounts of redeposited Brisi Limestone particles and lithoclasts in the younger Klaus and Rankweil Beds (uppermost Aptian to Lower Albian, respectively, to uppermost Albian), however, demonstrate that Brisi Limestone sediments once covered proximal areas of the Vorarlberg inner shelf. During a subsequent erosional phase at the Aptian-Albian boundary, these carbonate deposits were almost completely removed, transported, and redeposited on the distal inner and proximal outer shelf (Sect. 2.5.1).

2. In intermediate inner shelf areas, a several km-wide sandsheet developed (narrowing toward the east; Fig. 7; middle Upper Aptian **Brisi Beds**; max.

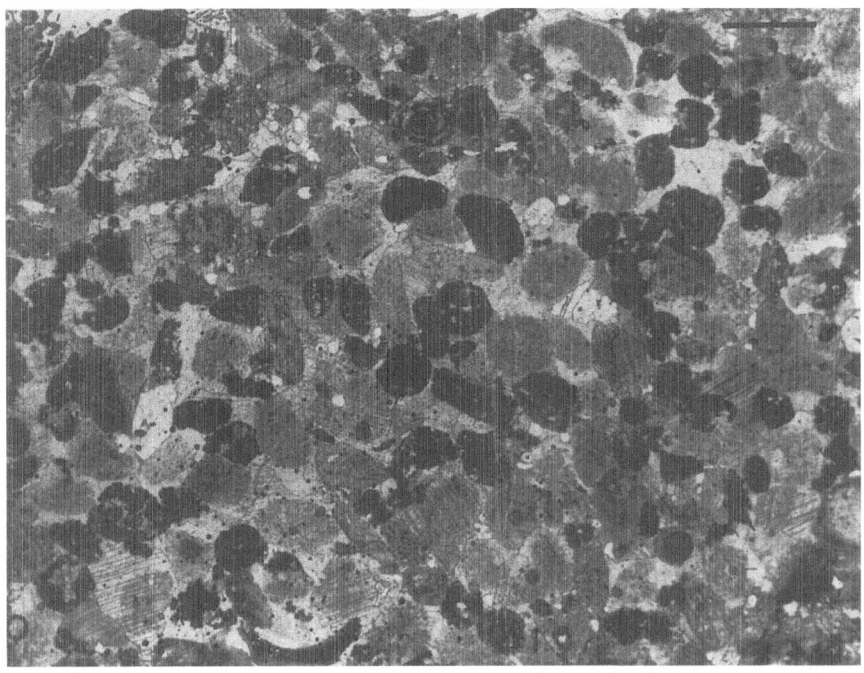

Fig. 8. Thin-section photomicrograph of Brisi Limestone sediments (<u>bar</u> = 1 mm). Note the abundance of crinoid fragments embedded in syntaxially grown sparites, and the presence of reworked, dark, and micritized Schrattenkalk particles (miliolids, orbitolinids), and lithoclasts

50 m thick; Fig. 2; sections 3-9 in Fig. 7). Brisi Beds include medium to coarse-grained (0.15-0.25 mm) glauconitic sandstones, which display distinct changes in facies types and thicknesses from proximal to distal areas (Fig. 7):

A. Proximal Brisi Bed sequences are thin (<10 m) and consist generally of a basal, tangentially cross-stratified bed (including over 4-m-long foresets; Fig. 9), overlain by a complex of tabular cross-stratified, and laminated beds, as well as small-scale tangentially, and probably hummocky cross-stratified beds (e.g., section 4 in Fig. 7).
B. Intermediate Brisi Bed sequences increase in thickness (20-50 m) and comprise irregularly bedded (2-20 cm thick) glauconitic sandstones (section 5 and top of sections 7 and 8 in Fig. 7).
C. Distal Brisi Bed sediments (20-30 m) consist of well-bedded (1-10 cm thick), bioturbated, and muddy sandstones (section 6 in Fig. 7), with sporadically intercalated beds of spiculitic pseudosparites (sections 8 and

Fig. 9. Tangentially cross-stratified, glauconitic sandstone bed at the base of the Brisi Beds (Bb; facies A), on top of the Schrattenkalk Formation (Skf; section is inverted). Note the approximately 4-m-long, well-individualized, internally plane-laminated foresets

9 in Fig. 7).

Facies B and C interfinger strongly and their lithologies appear in most cases as intimately interbedded packages within one section, indicating subtle variations in the supply of siliciclasts. Facies C, on the other hand, is transitional to the facies of the Gams Beds (see below).

Brisi Beds in intermediate and distal localities tend to coarsen upward (e.g., sections 7, 8, and 9 in Fig. 7). Upper portions in proximal Brisi Bed sequences include various amounts of Brisi Limestone particles. Proximal Gams Beds sequences are overlain by Brisi Beds (see below) and, in eastern and central Swiss localities, Brisi Beds are overlain by Brisi Limestone (Keller 1983; Ouwehand 1987). All are indicative of the southward progradation of the Brisi calcarenites and sands (Fig. 10).

In Vorarlberg, the presence of Brisi Beds sands, redeposited in younger Klaus and Rankweil Beds and an unconformity present at the top of Brisi

Fig. 10. Middle late Aptian.
Diagrammatic reconstruction perpendicular to the shelf. Note minima in net sediment accumulation rates in the proximal inner shelf area, along the drowning platform margin, and in the distal outer shelf area

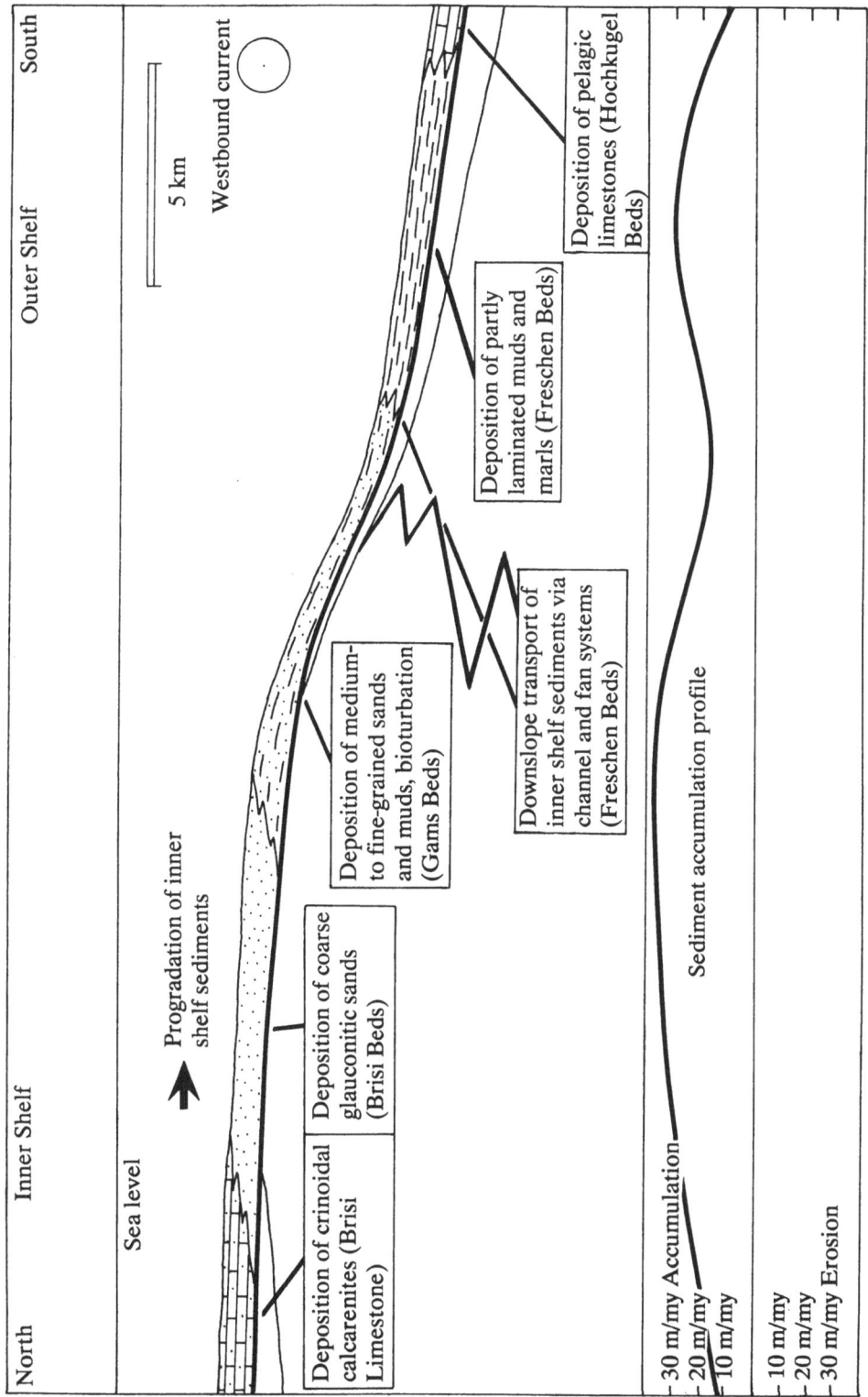

Bed sequences in proximal sections, suggest that the Brisi sandsheet once extended further northward. Proximal portions were remobilized, and redeposited in a subsequent, erosional episode at the Aptian-Albian boundary (simultaneously with Brisi Limestone calcarenites; see above and Sect. 2.5.1; Fig. 16).

3. Strongly bioturbated, dark, glauconitic, medium to fine-grained muddy sands blanketed the transitional area between the inner and outer shelf (0.1-0.25 mm; middle Upper Aptian **Gams Beds**; max. 30 m; Figs. 2, 7, and 10; sections 8-11 in Fig. 7). Gams Beds sporadically include beds of spiculitic micrites and pseudosparites, comparable to those in distal Brisi Beds (facies C). In the distribution area of the Gams Beds, isolated, mud-free sandridges of several 100-m width formed (section 11 in Fig. 7). In composition, sediments in these ridges are comparable to Brisi sands of facies type B.

4. Dark, commonly laminated and poorly bioturbated, *Hedbergella*-rich muds and marls accumulated on the outer shelf (**Freschen Beds**; max. 50 m; lower sequence = middle Upper Aptian to Lower Albian; upper sequence = Lower Albian to Upper Albian; Fig. 2, sections 12-18 in Fig. 7). The lower sequence of the Freschen Beds in proximal outer shelf areas include turbiditic event beds, which consist of silici- and bioclasts derived from the inner shelf. The assemblage of turbidites is assigned to channel and fan systems (Fig. 11):

A. **Channel system**: stacks of laterally discontinuous, coarse-grained, and normally graded Bouma-T_A deposits, bounded by sequences with isolated Bouma-T_A deposits are evidence of shifting channels, thus indicating a gentle dipping slope between the inner and outer shelf (Figs. 10 and 11).

B. **Proximal fan system**: sequences of thin (5-20 cm), medium to fine-grained, normally graded, and bioturbated muddy sandstones, as well as bioclasts (concentrated in calciturbidites; e.g., section 15 in Fig. 7) probably represent transitional zones between the channel and fan systems. Intercalated beds of bioturbated mudstones, including very fine-grained siliciclastic detritus, may represent overbank deposits (Fig. 11).

C. **Mid-fan system**: sequences consist of mm to cm-thick, regular layers of fine-grained, plane or ripple cross-laminated, locally slump-folded siliciclasts (Bouma-T_{B-C}; section 17 in Fig. 7; Fig. 12). Well-preserved trace fossils on turbidite bedsoles are characteristic for this depositional environment and may embody a syndepositional, probably imported sediment-feeding endofauna (*Scolicia* sp., *Zoophycus* sp.; Fig. 13).

D. Since most of the terrigenous clays in the "background" Freschen Beds,

Fig. 11. Schematic diagram of the morphology of the proximal outer shelf channel and fan systems within the lower portion of the Freschen Beds (middle Upper Aptian)

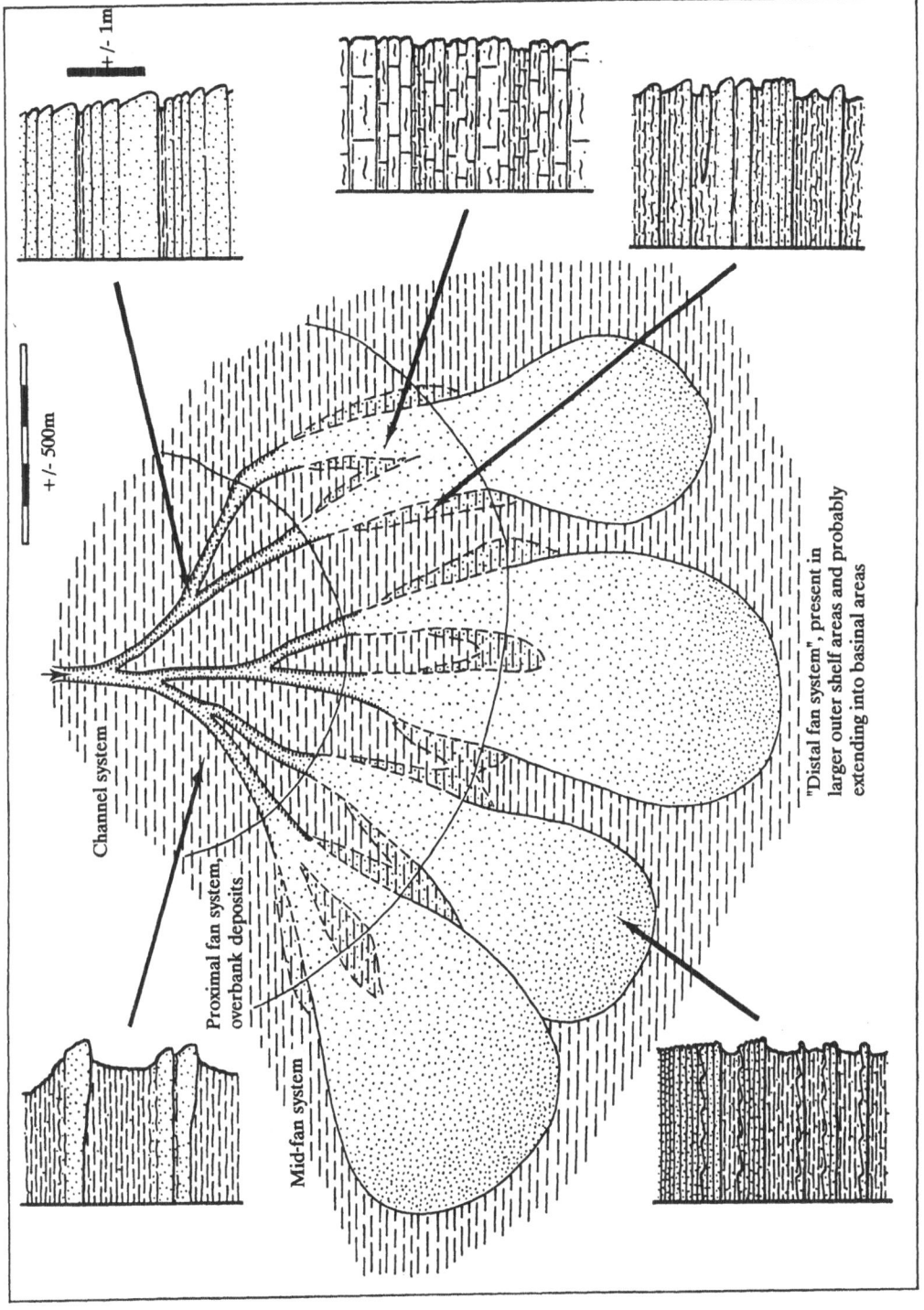

+/- 1m

+/- 500m

Channel system

Proximal fan system, overbank deposits

Mid-fan system

"Distal fan system", present in larger outer shelf areas and probably extending into basinal areas

Fig. 12. Mid-fan system, Freschen Beds (middle Upper Aptian).
Asymmetric current ripples in section <u>17</u> in Fig. 7 (approximately southward directed current flow; coin diameter = 2.5 cm). Interpreted as Bouma-T_C sequence, deposited on mid-fan lobes

as well as the very fine-grained, calcareous particles within the Hochkugel Beds (see below) are considered to have been derived *via* such channel and fan systems, the **"distal fan system"** (characterized by Bouma-T_{D-E} sequences) appears to have covered major areas of the outer shelf and probably parts of the basinal, Penninic area beyond it (*schistes lustres*; Pantic and Gansser 1977; Pantic and Isler 1978; Probst 1980).

5. Outcrops in the area of the "Hohe Kugel" (in a small, klippen-like nappe remnant of probably ultrahelvetic origin; Fig. 1; Oberhauser 1953, 1958, 1982; Föllmi 1986) display sediments of the Garschella and Seewen Formations in a facies characteristic of distal outer shelf regions. The lower sequence of the Freschen Beds is replaced by an alternation of spiculitic, radiolarian, and *Hedbergella*-bearing micrites and pseudosparites (comparable to calcareous beds in Brisi Beds, facies type D, and in Gams Beds), and dark, laminated, Freschen Beds-like marls and clays (Upper Aptian **Hochkugel Beds**; max. 15 m; section 18 in Fig. 7). The carbonate beds include mm-thin, partly burrowed laminae of very fine-grained (<0.05 mm), silici-

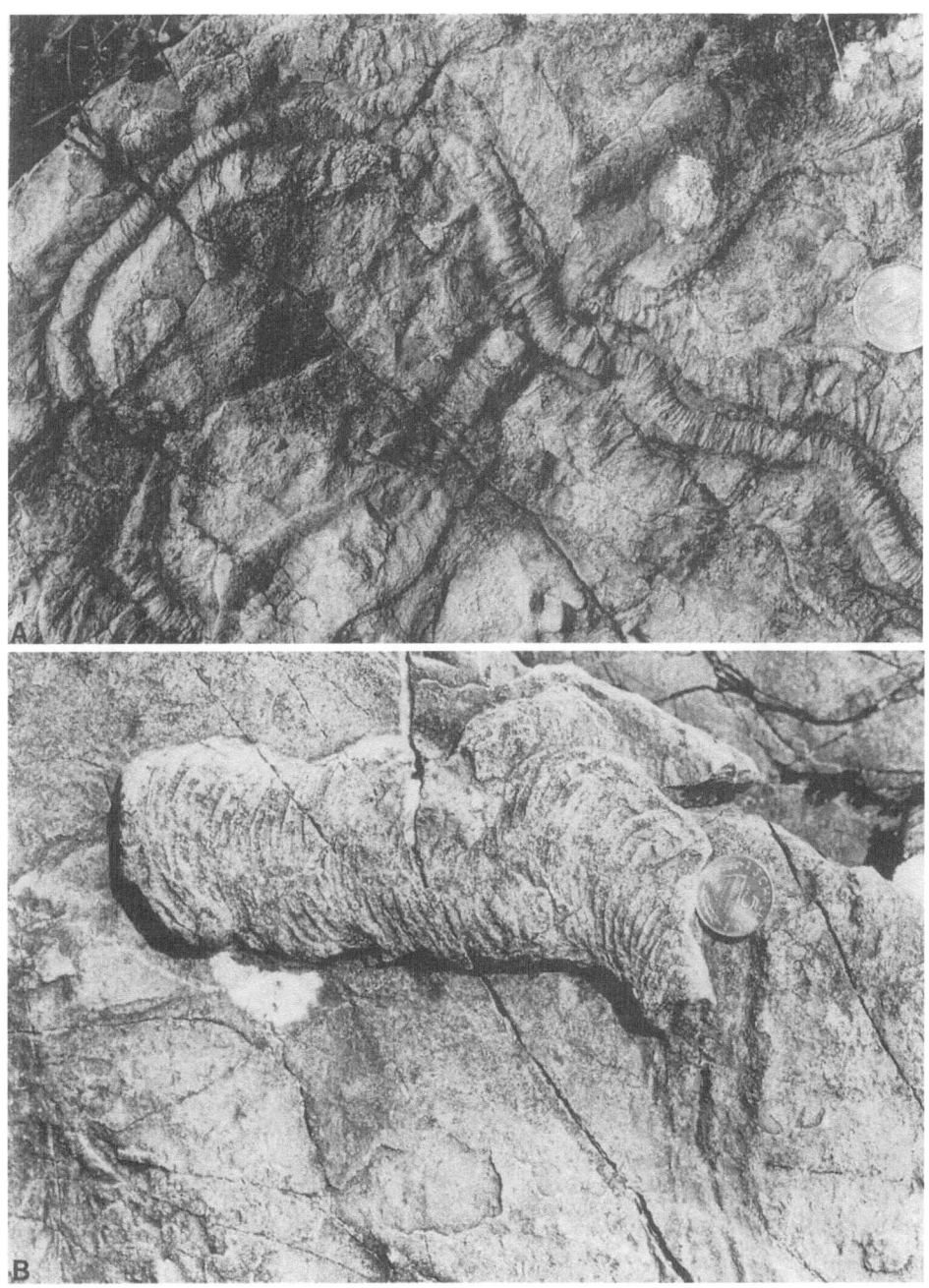

Fig. 13. Mid-fan system, Freschen Beds (middle Upper Aptian).
A. *Scolicia* sp. and B. *Zoophycus* sp., both preserved on bedsoles of thin-bed-ded Bouma-T_{C-D} sequences in section 1̲7̲ of Fig. 7 (coin diameter = 2.5 cm)

Fig. 14. Hochkugel Beds.
A. Slump folds in Hochkugel Beds (W Hohe Kugel). Note the rapid lateral changes in thicknesses of the carbonate beds. Hammer in centre for scale.
B. Thin-section photomicrograph (<u>bar</u> = 1 mm). Approximately 1 cm thick lamina of very fine-grained (<0.05 mm) pellets, debris of benthic foraminifera, and siliciclastic particles, within spiculitic, radiolarian micrites. Interpreted as deposition from a contourite current

clastic, and calcareous particles such as pellets and debris of benthic foraminifera (Fig. 14B). These distinct laminae and the interbedded laminated
marls and clays may represent contourites; i.e., sediments, resuspended and
redeposited by bottom currents, derived originally within turbidite flows
from the above discussed channel and fan systems ("distal fan system"). Indicative of this are the abundance and regular distribution of such redeposited sediments in the Hochkugel Beds, the irregular and lenticular bedding
within the Hochkugel Beds, probably evoked by gentle bottom currents
(Fig. 14A), and the widespread facies of laminated clays throughout the
outer shelf region (in Freschen Beds and intercalated in Hochkugel Beds),
probably distributed by bottom currents (Fig. 11).

The typical Hochkugel Beds facies is restricted to the ultrahelvetic Hohe
Kugel area. A facies transitional to the lower sequence of the Freschen
Beds is preserved in section 16 (Fig. 7).

2.4.2 Time-Space Relations

Brisi Limestone, Brisi Beds, Gams Beds, the lower sequence of Freschen Beds,
and Hochkugel Beds each characterize a facies zone, roughly parallel to the
shelf break and bounded by typically interfingering facies transitions (prograding southward on the inner shelf; Figs. 7 and 10). The sequences are approximately coeval and constitute a single depositional system tract (Haq et al.
1987).

Age dating within this depositional system is difficult, due to the scarcity
of guide fossils. Age constraints on the inner shelf part are derived from un-
der- and overlying phosphatic beds (youngest Luitere Bed ammonoids: *melchioris* Zone, oldest ammonoids in Rankweil Beds: *jacobi* Zone; Sect. 2.5.2).
The Brisi Limestone, Brisi Beds, and Gams Beds are essentially of *nolani/
nodosocostatum* Zone age (and portions of the *melchioris* and *jacobi* Zones).
The outer shelf part is underlain by the Lower to lower Upper Aptian Mittagspitz Formation (Sect. 2.3.2) and overlain by the uppermost Aptian and Albian Rankweil Beds and Freschen Beds (Sects. 2.5.2 and 2.6.2), which similarly
suggests a middle late Aptian age.

The middle Upper Aptian detritus-rich depositional system tract probably
developed within 1 my.

2.5 Latest Aptian to Earliest Albian

2.5.1 General Overview

In latest Aptian time (*jacobi* Zone), the deposition of predominantly detritus-
rich, clastic sediments ended, and a regime of retarded sediment accumulation
took over, dominated by major episodes of erosion and redeposition (Figs. 15
and 16).

1. Sediments of this time interval are absent in the proximal inner shelf sections of Vorarlberg. A prominent discontinuity separates Brisi Beds or Schrattenkalk sediments from distinctly younger Albian sediments (section 3 in Fig. 15; Fig. 16).
   In eastern Swiss proximal inner shelf sequences, a phosphatic bed of late Aptian and earliest Albian age is preserved on top of the Brisi Limestone (upper Upper Aptian to lowermost Albian **Twäriberg Bed**; max. 0.7 m; Fig. 16; section 1 in Fig. 15 = section CV in Ouwehand 1987; cf. also Föllmi and Ouwehand 1987, Fig. 5). Typical Twäriberg Bed sequences are built up of thin, glauconite-rich, muddy sandstones, which cover and infill the phosphatized, rugged, surface of the uppermost Brisi Limestone bed. Especially within the surface fissures, various amounts of phosphatic particles are preserved (fossil debris and lithoclasts).
2. In distal areas of the Vorarlberg inner shelf, a thin cover of heterogeneous, allochthonous sediments accumulated (uppermost Aptian to lowermost Albian **Klaus Beds**; generally <1 m, max. 5-7 m; Fig. 2; sections 4-9 in Fig. 15; Fig. 16). Klaus Beds consist of glauconitic sandstones, which include various amounts of exotic particles (5-75%; Figs. 17 and 18):

   A. Crinoid, bryozoan, bivalve, and algal fragments, micritized Schrattenkalk litho- and bioclasts, as well as darkened pebbles. The spectrum of composition, size, relative abundance, and appearance of these particles is identical to that of Brisi Limestone particles.
   B. Millimeter to 1-cm-sized, variegated quartz particles, typical of Klaus and Rankweil Beds (see below).
   C. Phosphatized fossil debris (commonly with multiple accreted phosphate generations) and phosphatized lithoclasts of the Brisi Limestone (ranging from granule to boulder size; Fig. 18B) and the Schrattenkalk Formation (ranging from granule to pebble size). The age of present phosphatized ammonoids corresponds to the age of Twäriberg Bed ammonoids.
   D. Micritic granules and pebbles, generally lacking macro- and microfossils, with the occasional exception of packed shell debris of epibiontic bivalves (*Aucellina* sp.).

The presence of exotic particles (not correlatable to the sediments from immediately underlying beds), shallow incised channels (1-5 m wide, gene-

Fig. 15. Latest Aptian to earliest Albian.
Palinspastic map indicates the inner shelf areas dominated by erosion, the position of the Klaus Beds (distal inner shelf), the position of observed channels infilled with Rankweil Beds (along the Rankweil Ramp = slope between inner and outer shelf), as well as the distribution of the Freschen Beds. Correlation of uppermost Aptian to lowermost Albian beds is indicated by dots. Note different scale for section 4 (1-m scale bar for section 4; 2-m scale bar for sections 1-3 and 5-10)

31

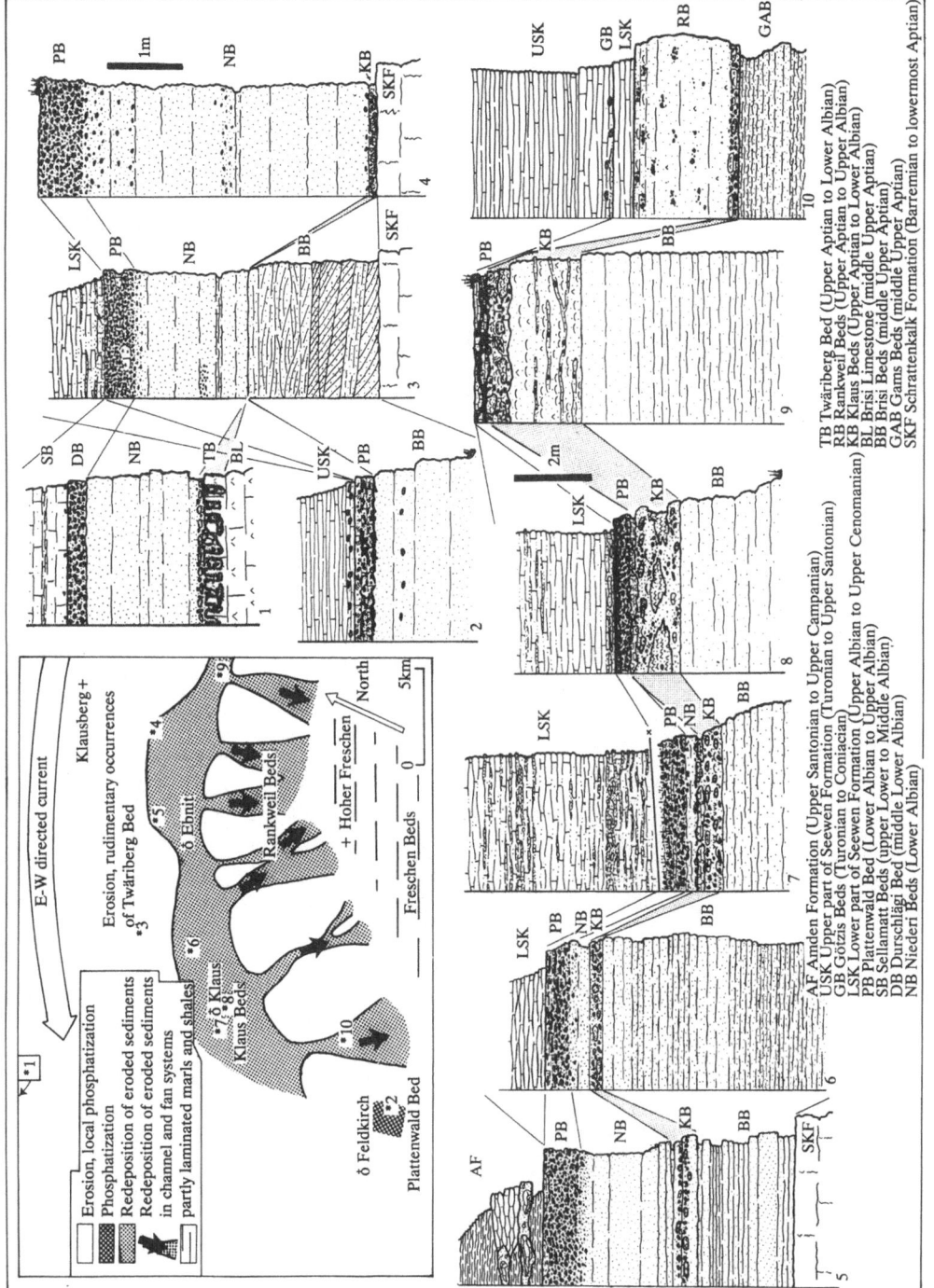

E-W directed current

Klausberg +

Erosion, rudimentary occurrences
of Twäriberg Bed *3

Erosion, local phosphatization

Phosphatization

Redeposition of eroded sediments

Redeposition of eroded sediments
in channel and fan systems

partly laminated marls and shales

*1

North

5km

0

δ Ebnit

*4

*5

*2

Rankweil
Beds

Hoher Freschen

Freschen Beds

*6

*7 δ Klaus
*8 Klaus Beds

δ Feldkirch
*2

*10

Plattenwald Bed

AF   Amden Formation (Upper Santonian to Upper Campanian)
USK Upper part of Seewen Formation (Turonian to Upper Santonian)
GB Götzis Beds (Turonian to Coniacian)
LSK Lower part of Seewen Formation (Upper Albian to Upper Cenomanian)
PB Plattenwald Bed (Lower Albian to Upper Albian)
SB Sellamatt Beds (upper Lower to Middle Albian)
DB Durchschlägi Bed (middle Lower Albian)
NB Niederi Beds (Lower Albian)

TB Twäriberg Bed (Upper Aptian to Lower Albian)
RB Rankweil Beds (Upper Aptian to Upper Albian)
KB Klaus Beds (Upper Aptian to Lower Albian)
BL Brisi Limestone (middle Upper Aptian)
BB Brisi Beds (middle Upper Aptian)
GAB Gams Beds (middle Upper Aptian)
SKF Schrattenkalk Formation (Barremian to lowermost Aptian)

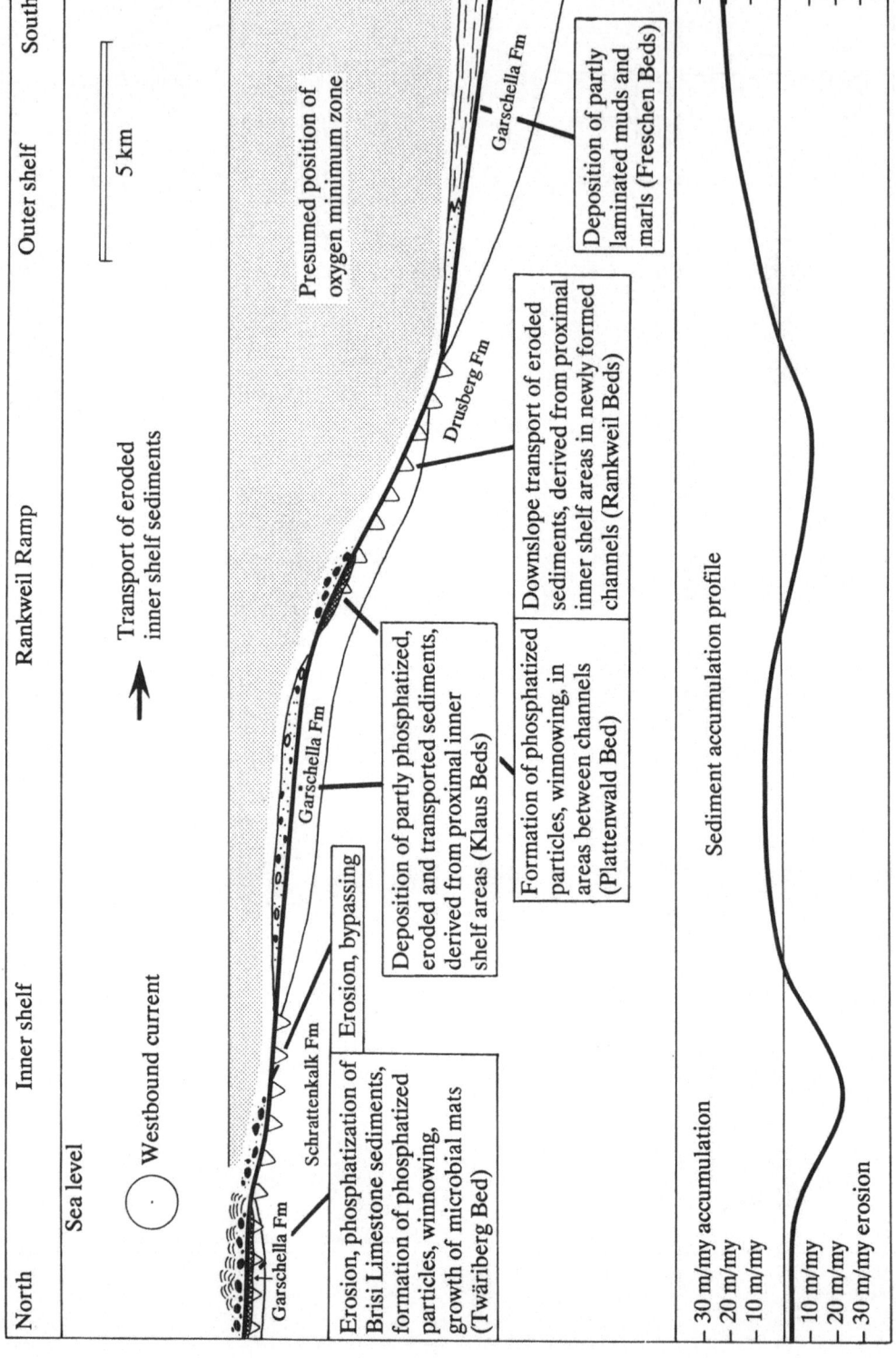

North — Inner shelf — Rankweil Ramp — Outer shelf — South

Sea level

Westbound current

Transport of eroded inner shelf sediments

5 km

Presumed position of oxygen minimum zone

Garschella Fm

Schrattenkalk Fm

Garschella Fm

Drusberg Fm

Garschella Fm

Garschella Fm

Erosion, phosphatization of Brisi Limestone sediments, formation of phosphatized particles, winnowing, growth of microbial mats (Twäriberg Bed)

Erosion, bypassing

Deposition of partly phosphatized, eroded and transported sediments, derived from proximal inner shelf areas (Klaus Beds)

Formation of phosphatized particles, winnowing, in areas between channels (Plattenwald Bed)

Downslope transport of eroded sediments, derived from proximal inner shelf areas in newly formed channels (Rankweil Beds)

Deposition of partly laminated muds and marls (Freschen Beds)

Sediment accumulation profile

30 m/my accumulation
20 m/my
10 m/my

10 m/my
20 m/my
30 m/my erosion

rally <1 m deep, either at the base or within the Klaus Beds), rapid lateral facies changes, and a discontinuity at the base, indicates that the Klaus Beds essentially consist of allochthonous, i.e., eroded, transported, and redeposited sediments, which were derived from proximal inner shelf areas (from exposed sediments of the Twäriberg Bed, Brisi Limestone, Brisi Beds, and Schrattenkalk Formation; Föllmi 1986; Föllmi and Ouwehand 1987). The redeposited sediments accumulated in channelized and sheet-like debris flow deposits (debrites).

The lithologic character of the redeposited sediments provide valuable insight into the composition of sediments on the Vorarlberg proximal inner shelf prior to and during this erosive phase (Fig. 17). Prior to the erosive episode, proximal areas of the inner shelf were covered with Brisi Limestone sediments, more distal areas with Brisi Beds (in analogy to the distribution of *in situ* preserved Brisi Limestone and Beds in eastern Switzerland; Ganz 1912; Heim 1910-1917; Ouwehand 1987). In episodic events (storms?), upper portions of the Brisi Limestone sediments were remobilized, transported seaward, and incorporated into Brisi Beds (phase 1 in Fig. 17; middle late Aptian time). During the erosive episode, Brisi Limestone and Brisi Bed sediments, present in the proximal inner shelf area, experienced sequential removal, until the surface of the Schrattenkalk Formation was reached (smaller amounts of the Schrattenkalk top beds were eroded as well). During intervening, more "quiet" periods, autochthonous and condensed phosphatic sediments developed on top of the erosional surface (penetrative phosphatization of Brisi and Schrattenkalk carbonates along the periphery of the truncated erosional surface and formation of phosphatic particles: Twäriberg Bed) and micritic oozes accumulated (phases 2-4 in Fig. 17; latest Aptian and earliest Albian interval).

The truncated upper surface of the Brisi Limestone, as well as the Brisi Limestone pebbles redeposited into the Klaus Beds, are indicative of a progressive, early diagenetic lithification of Brisi Limestone, prior to this episode of erosion and phosphatization.

3. Perpendicular to the boundary area between the inner and outer shelf, a new set of channels formed, cut into or through middle Upper Aptian Gams Beds (palinspastic map in Fig. 15 shows locations of the known channels). These channels reached a width of several 100 m, eroded up to 25 m deep (amount of eroded, missing Gams Beds, when compared to adjacent, uneroded Gams Beds; e.g., sections 9 or 10 in Fig. 7), and reached a length of 5-7 km. Channel infills consist partly of redeposited sediments, similar to and coeval with Klaus Beds, partly of younger redeposited sedi-

Fig. 16. Latest Aptian to earliest Albian.
Diagrammatic reconstruction perpendicular to the shelf. Note the inner shelf area of phosphatization and erosion of underlying sediments (Brisi Limestone and Brisi Beds) and the new formation of prominent channels along the steepened ramp between the inner and outer shelf

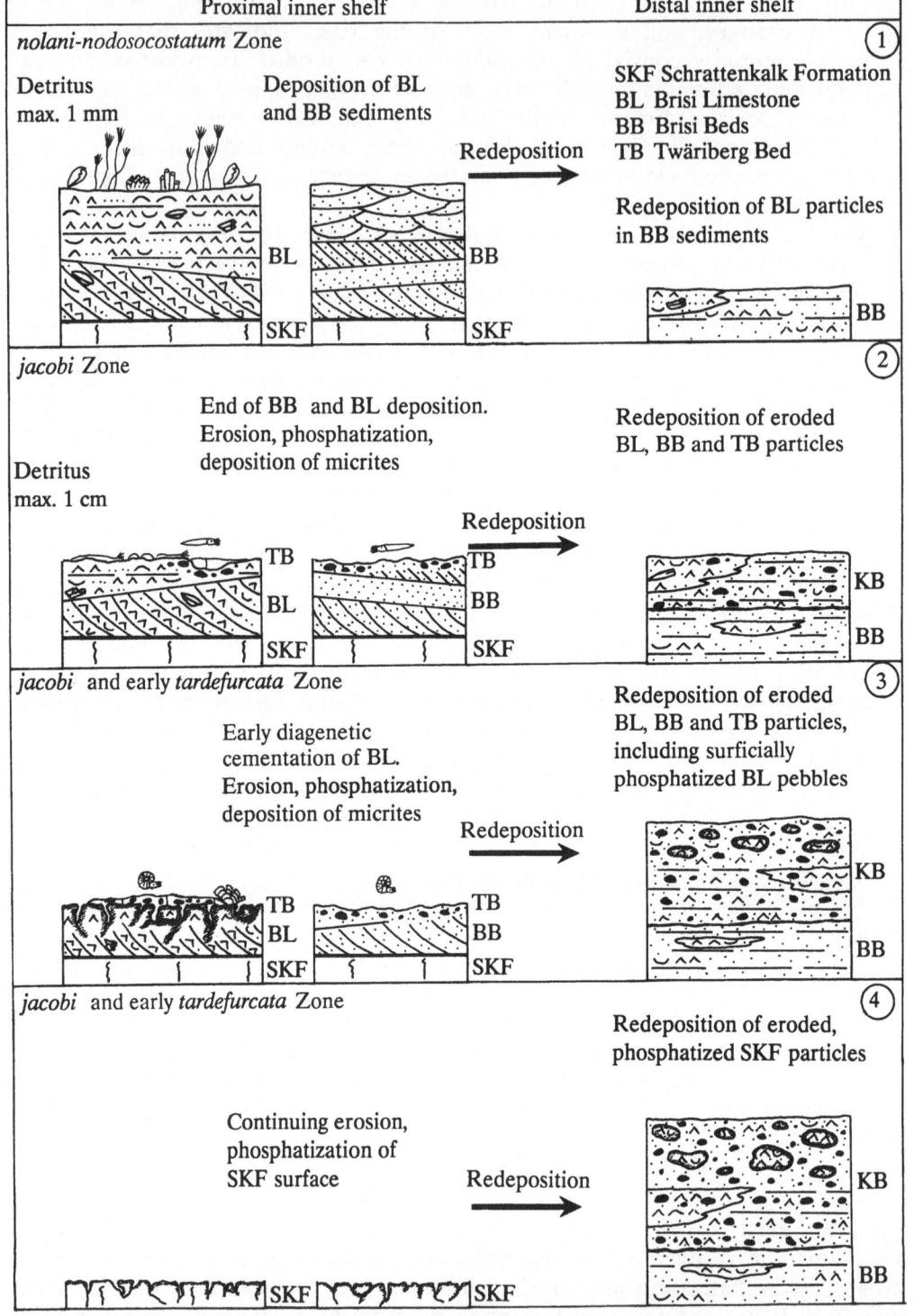

ments (up to uppermost Albian in age; see below). They are grouped together into the uppermost Aptian to uppermost Albian **Rankweil Beds** (max. 25 m; Fig. 2, section 10 in Fig. 15, sections 14 and 15 in Fig. 19). The presence of erosive bases (with flute and load casts), basal conglomerates, and normally graded and laminated beds within the Rankweil Beds point to gravity flow deposition, in most cases as turbidites.

These newly formed channels are more proximal and much larger than the above described middle Upper Aptian channel and fan systems. They remained in a stable position and were in most cases actively infilled throughout the entire Albian (Figs. 15 and 19). The channel dimensions and stability probably had their origin in a process of steepening (and faulting?) of the transitional area between the inner and outer shelf (= "Rankweil Ramp"), due to differential subsidence of the inner and outer shelf (Sect. 3.2.2).

Intriguingly, these incised channels formed during a (relative) sea level rise, unlike most modern canyons (Sects. 3.2.2 and 3.7; e.g., Moslow et al. 1988).

4. Along the Rankweil Ramp, the areas between the channels were not affected by the erosive episode and influenced by the influx of Klaus or Rankweil redeposited sediments. The interchannel areas are characterized by persistent low sediment accumulation rates (during latest Aptian to earliest Turonian) and by extensive phosphatization, which resulted in the formation of autochthonous and condensed phosphates (during latest Aptian to earliest Albian; section 2 in Fig. 15; Figs. 16 and 19; Plattenwald Bed; Sect. 2.6.1).

5. In outer shelf areas, sedimentation of the lower sequence of the Freschen Beds continued. Proximal sections include beds of coarse-grained, redeposited sediments, comparable to sediments of the Klaus Beds (e.g., section 12 in Fig. 7). These beds depict distal offshoots of the Rankweil Beds (Fig. 16).

2.5.2 Time-Space Relations

The Vorarlberg inner shelf part of this depositional system tract consists almost entirely of older, reworked sediments. An exception is given with the micritic and phosphatized particles and nodules, included in the Klaus and Rankweil Beds, which are considered to be formed and subsequently eroded during the build-up of this sequence.

Fig. 17. Schematic reconstruction, illustrative of the sequential deposition of the Klaus Beds (KB).
1. Situation in middle late Aptian time, during formation of the Brisi Limestone and Brisi Beds, before deposition of the Klaus Beds.
2-4. Situation in latest Aptian and earliest Albian times, during deposition of the Klaus Beds

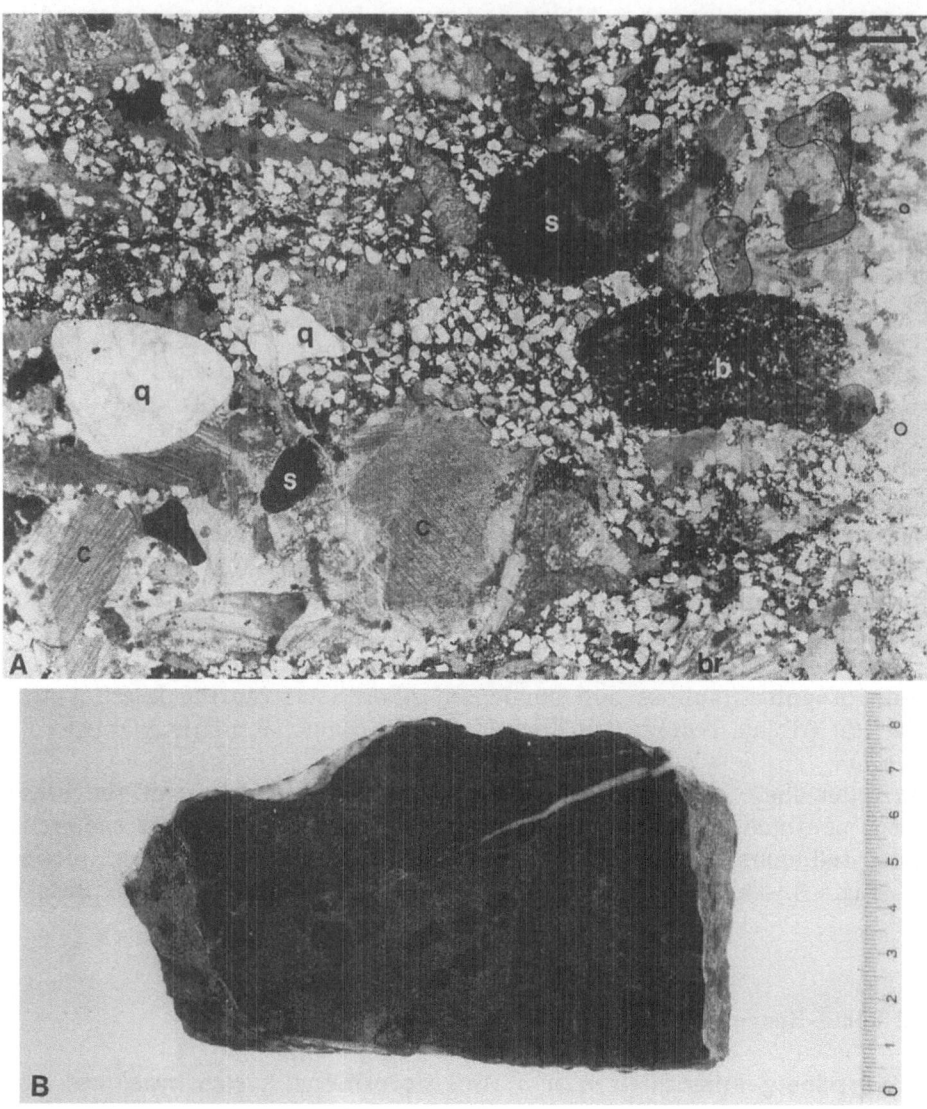

Fig. 18. Klaus Beds.

A. Characteristic thin-section photomicrograph (bar = 1 mm) of the Klaus Beds (phase 2 in Fig. 17). Crinoid fragments (c), bryozoa debris (br), micritized Schrattenkalk particles (s), and darkened pebble (b), all derived from eroded Brisi Limestone; large quartz particles (q), characteristic for the Klaus Beds and coeval sediments within Rankweil Beds; phosphatized particles, derived from Twäriberg Bed. Matrix consists of glauconitic sand and is probably derived from Brisi Beds.

B. Slabbed specimen (phase 3 and 4 in Fig. 17). Dm-sized, surficially phosphatized Brisi Limestone pebbles and cm-sized phosphatic diaclasts (derived from Twäriberg Bed), embedded in glauconitic sands (probably derived from Brisi Beds)

The presence of the phosphatized ammonoids *Hypacanthoplites* and *Beu-danticeras* (*Pseudorbulites*) in the Klaus Beds and at the base of the Rankweil Beds, the presence of *Hypacanthoplites, Leymeriella* (*Leym.*) and *Beudanticeras* (*Pseudorbulites*) in phosphatic beds of the interchannel areas and of *Leymeri-ella* (*Neoleym.*) in the Plattenwald Bed, superjacent to Klaus Beds, indicate the *jacobi* and the early *tardefurcata* Zones as the time range in which this depositional sequence developed. This age is consistent with the biostratigraphic results obtained from east Swiss Twäriberg Bed localities (presence of *Hypacanthoplites* and *Leymeriella*; Ganz 1912; Föllmi and Ouwehand 1987).

2.6 Early to Late Albian

2.6.1 General Overview

Albian sediments of the Vorarlberg inner shelf reflect an episode of extensive phosphatization and condensation. In proximal and intermediate inner shelf areas, general processes of low sediment accumulation were punctuated by several intervals of increased detritus input (Niederi, Sellamatt, and Aubrig Beds), whereas distal inner shelf sediments indicate persistent condensation throughout the period from latest Aptian to latest Albian, locally to late Cenomanian and early Turonian. These "ultra"-condensed sediments document a complex history of deposition during 16 to 23 my (after Harland et al. 1982), commonly recorded in less than 50 cm sediment (e.g., section 13 in Fig. 19; Fig. 28).

1. Development of elongated sandbodies in the proximal and middle inner shelf:

   A. **Niederi Beds** (middle part of the *tardefurcata* Zone; max. 6 m; Fig. 2, sections 3-5 and 7-10 in Fig. 19) consist of poorly and irregularly bedded, fine-grained (0.05-0.1 mm), glauconitic, spiculitic, and thoroughly bioturbated sandstones (e.g., *Palaeophycus* sp., *Chondrites* sp.), and include occasionally lenses of *Hedbergella*-rich micrites (e.g., section 9 in Fig. 19).
   The Niederi Beds build up a sandbody that is oriented approximately parallel to the Rankweil Ramp (Fig. 19) and exhibits a strongly asymmetrical shape in cross-section perpendicular to the shelf (short, steep, proximal limb and elongated distal limb; Fig. 20). North of the Niederi Beds sandbody, younger sediments cover Schrattenkalk sediments, indicating nondeposition or subsequent erosion of the Niederi Beds (sections 1, 2, and 6 in Fig. 19). Time-equivalent sediments south of the Niederi sandbody were subjected to condensation processes, and are included in the Plattenwald Bed. Niederi Beds coarsen toward the top of the sandbody, as well as toward the thin veneers of sand at the distal, southward periphery of the sandbody. The asymmetric sandbody geometry is here interpreted as the result of "sandwiching" by a strong, erosive cur-

38

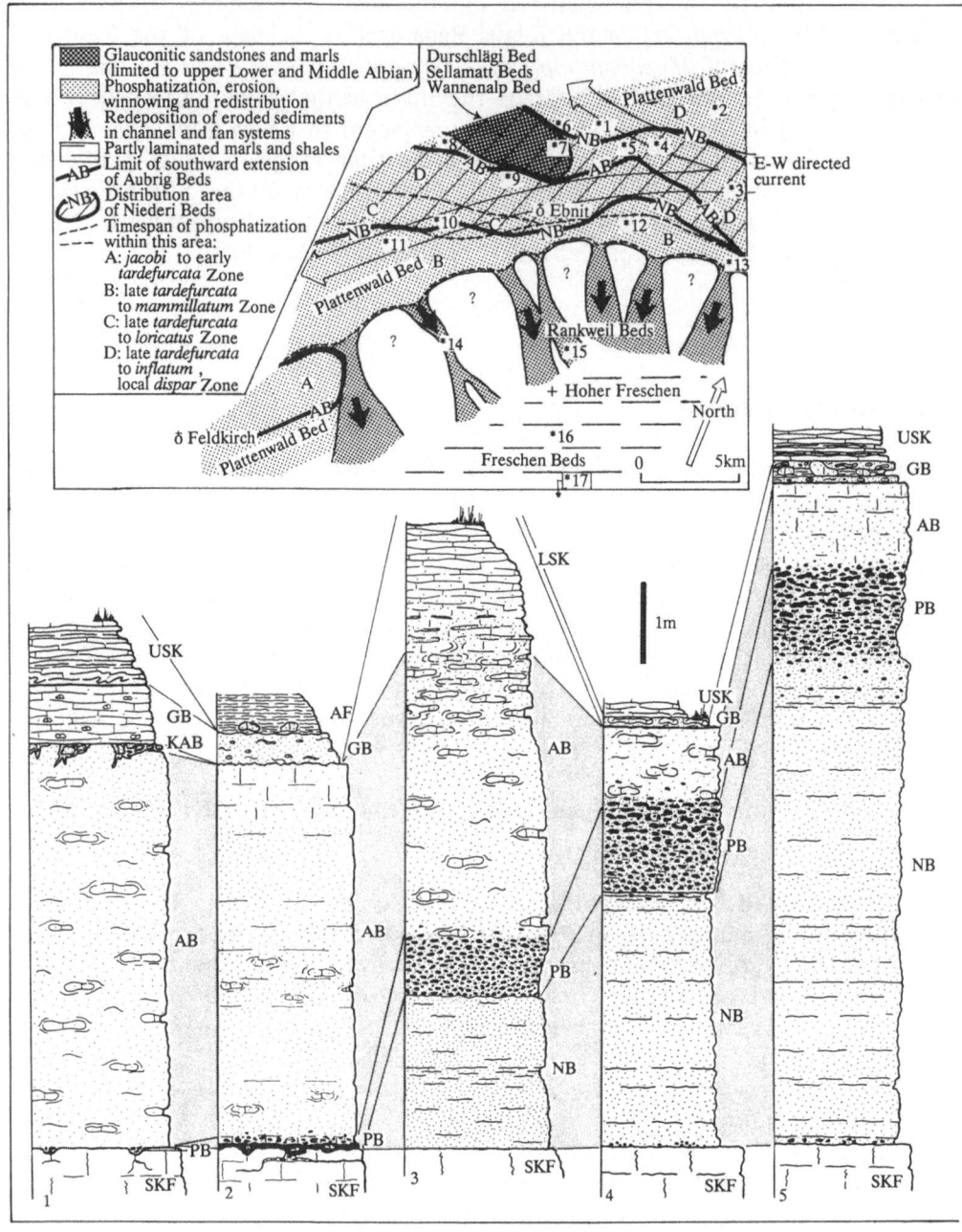

Fig. 19. Early to late Albian.
Palinspastic map displays distribution areas of the Durschlägi, Sellamatt, and
Wannenalp Beds (within same area), the Niederi, Plattenwald, Rankweil, and
Freschen Beds, as well as the southward extension of the Aubrig Beds. It
shows the areas of equal duration of phosphogenesis (T_{PH}) within the Plat-
tenwald Bed, as indicated by phosphatized ammonoids and inoceramids (zones

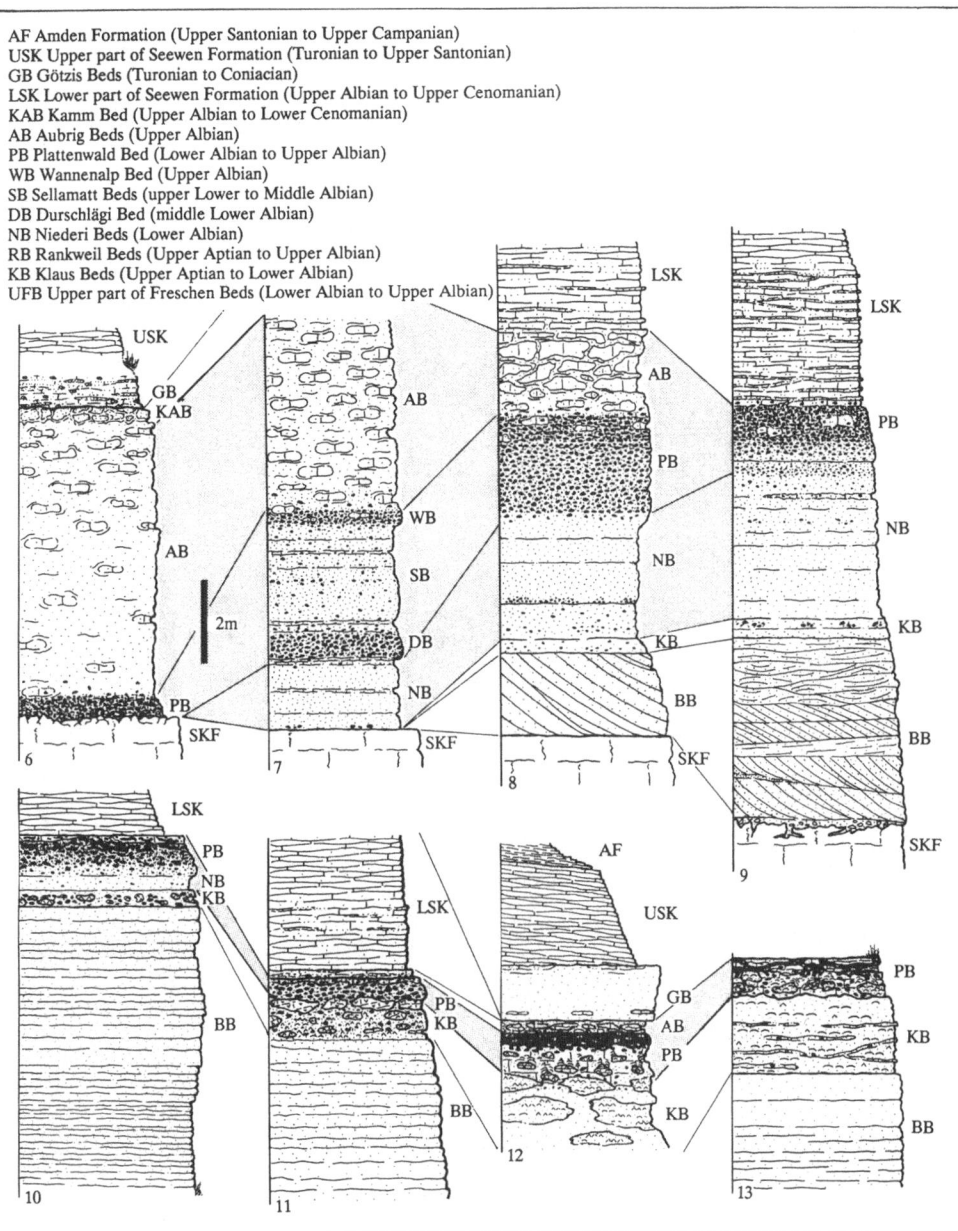

AF Amden Formation (Upper Santonian to Upper Campanian)
USK Upper part of Seewen Formation (Turonian to Upper Santonian)
GB Götzis Beds (Turonian to Coniacian)
LSK Lower part of Seewen Formation (Upper Albian to Upper Cenomanian)
KAB Kamm Bed (Upper Albian to Lower Cenomanian)
AB Aubrig Beds (Upper Albian)
PB Plattenwald Bed (Lower Albian to Upper Albian)
WB Wannenalp Bed (Upper Albian)
SB Sellamatt Beds (upper Lower to Middle Albian)
DB Durschlägi Bed (middle Lower Albian)
NB Niederi Beds (Lower Albian)
RB Rankweil Beds (Upper Aptian to Upper Albian)
KB Klaus Beds (Upper Aptian to Lower Albian)
UFB Upper part of Freschen Beds (Lower Albian to Upper Albian)

A, B, C, and D; A = latest Aptian to earliest Albian; B = early Albian; C = early and middle Albian; D = early to late Albian). The position of the channels along the Rankweil Ramp is identical to their position at the Aptian-Albian boundary (Fig. 15). Questionmarks in most of the interdistributory areas refer to the unknown character of Albian sediments in this zone, due to a younger erosive phase in which these sediments have been obliterated (channel

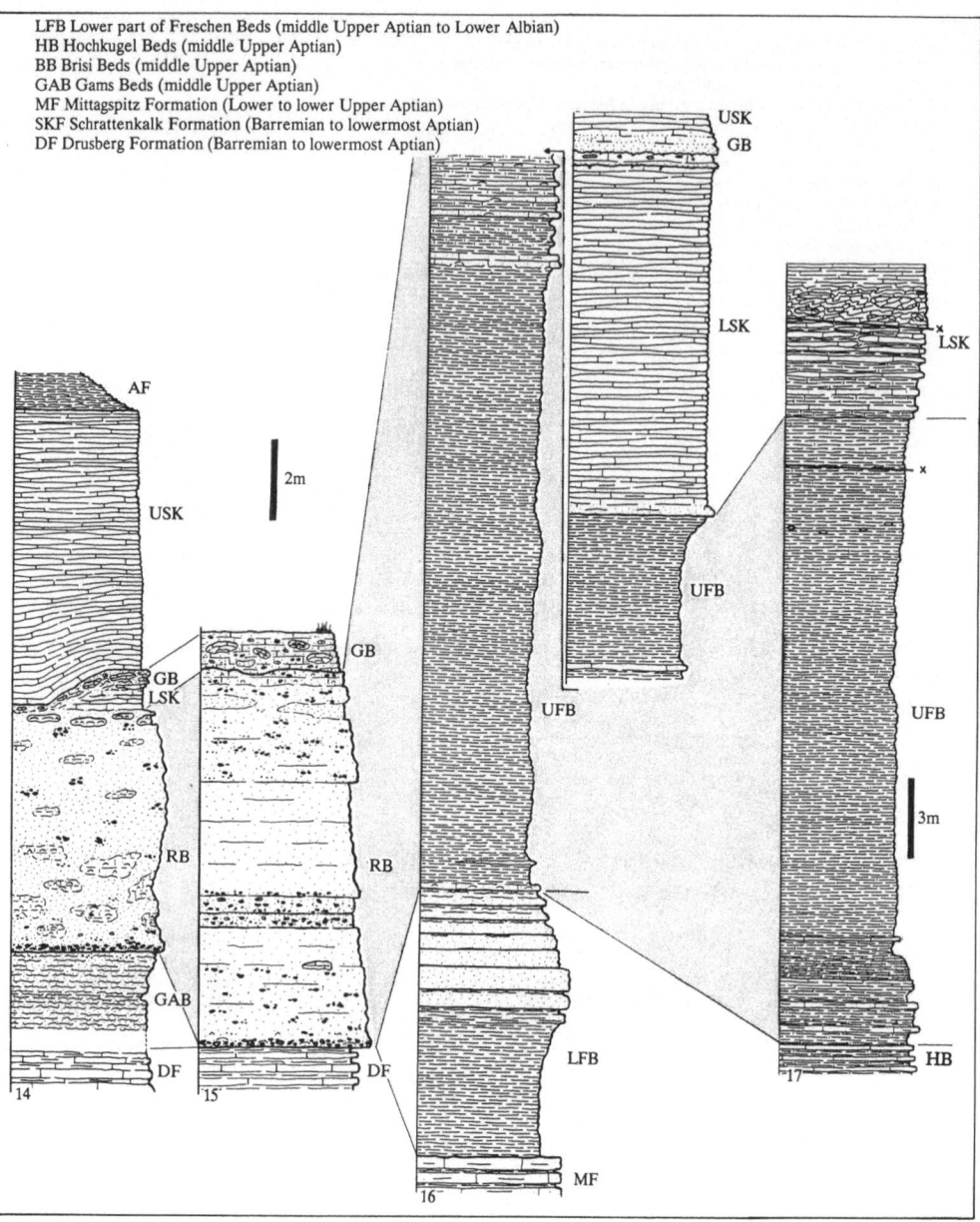

LFB Lower part of Freschen Beds (middle Upper Aptian to Lower Albian)
HB Hochkugel Beds (middle Upper Aptian)
BB Brisi Beds (middle Upper Aptian)
GAB Gams Beds (middle Upper Aptian)
MF Mittagspitz Formation (Lower to lower Upper Aptian)
SKF Schrattenkalk Formation (Barremian to lowermost Aptian)
DF Drusberg Formation (Barremian to lowermost Aptian)

Fig. 19 (continued).
sediments are preserved, because of their deep incision into middle Upper Aptian Gams Beds). Correlation of Lower to Upper Albian beds is indicated by dots. Note the different scales (1-m scale bar for sections 1-5; 2-m scale bar for sections 6-13; 3-m scale bar for section 17). Section 17 is located in the ultrahelvetic nappe of the Hohe Kugel

rent, which limited deposition of the Niederi Beds toward the north, and a less effective, winnowing current (deeper waters), which prevented the extension of the Niederi Beds sandbody to the south (Fig. 20). Increased particle size at the top of the sandbody reflects an increase in current strength or a shift of the current system (Sect. 3.4).

B. **Sellamatt Beds** (upper Lower to Middle Albian; max. 3 m; Fig. 2, section 7 in Fig. 19) consist of irregularly bedded, fine-grained (<0.1 mm) glauconitic sandstones and include minor amounts of phosphatized fossil debris (sponges, bivalves, gastropods, ammonoids, echinoids).

The Vorarlberg Sellamatt Beds form the eastern end of an elongated sandbody, approximately parallel to the Rankweil Ramp, which broadens and thickens rapidly toward the west, incorporating calcareous sediments and clays (e.g., *Hedbergella* micrites; Fig. 19; Ouwehand 1987). The Sellamatt sandbody is bounded northward, southward, and eastward by condensed phosphatic sediments (Plattenwald Bed; see below; Fig. 19). This wedge-shaped configuration probably resulted from current-induced "sandwiching", similar to the older Niederi sandbody. In the case of the Sellamatt sandbody, the current bifurcation point may have been located in the eastern part of the Vorarlberg inner shelf, as is indicated by the eastern end of the sandbody; in the case of the older Niederi Beds, this point was probably located in a more easterly direction (Fig. 19; Sect. 3.4).

C. **Aubrig Beds** (Upper *inflatum* and Lower *dispar* Zones; max. 1 m; Figs. 2 and 21; sections 1-8 in Fig. 19) comprise massive, medium to fine-grained (0.05-0.15 mm) glauconitic sandstones, which include various amounts of calcareous fossil debris (sponge spicules, benthic and planktonic foraminifera, inoceramid prisms, and echinoderm fragments). The presence of frequent calcareous, nodular structures bounded by bed-parallel stylolites, an uneven distribution of calcareous bioclasts in the nodules and in the surrounding sandy, glauconitic matrix, as well as the presence of neomorphic calcite (pseudosparite) within the nodules point to comprehensive $CaCO_3$ remobilization after deposition, probably due to diagenetic overprinting of intercalated $CaCO_3$-rich and $CaCO_3$-poor beds by selective pressure solution (Fig. 22).

The Aubrig Beds build up an extensive sandsheet on the proximal inner shelf, of which only the southward limit is known (more than 50 km width; Ouwehand 1987). Within this sandsheet, the Aubrig Beds coarsen upward (Fig. 21). Granulometric and biostratigraphic comparison of well-developed proximal sections with thin, distal sections indicate that distal Aubrig Beds have been subjected to progressively stronger condensation (Fig. 21). Time-equivalent sediments south of the Aubrig sandsheet are included in the strongly condensed Plattenwald Bed (Fig. 2; see below).

2. Condensation and phosphatization on the inner shelf:
The repeated influx of large amounts of siliciclastic detritus in proximal and middle parts of the inner shelf interrupted processes of phosphatization

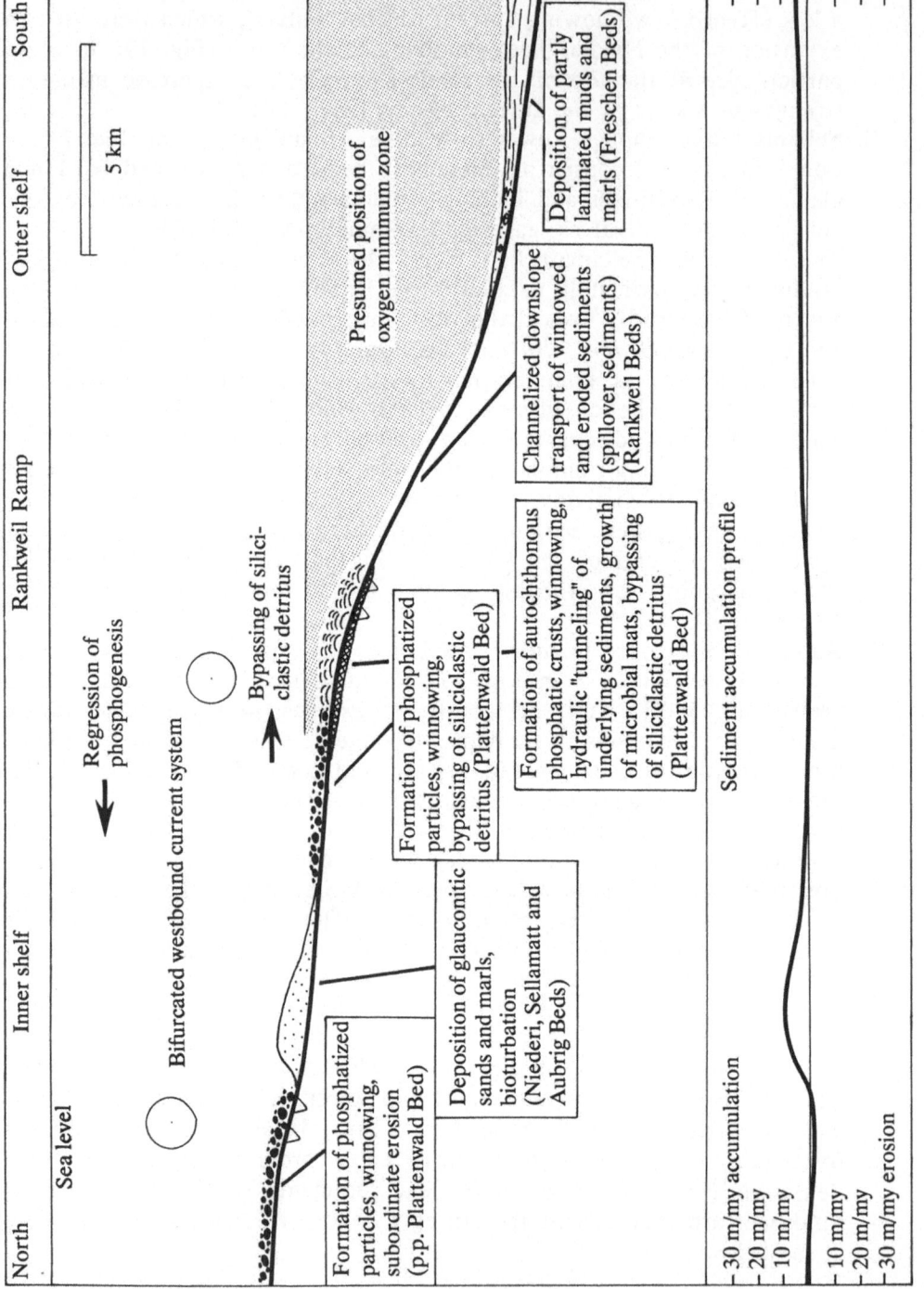

Fig. 20. Early to early late Albian.
Diagrammatic reconstruction perpendicular to the shelf. Note the asymmetrical shape of the middle inner shelf sandbody, induced by a bifurcated current system, and the overall low sediment accumulation rates. During the Albian period, the zone of phosphogenesis sequentially moved landward

and condensation and limited the formation of well-developed phosphatic sediments on the proximal inner shelf to the middle Albian ("basal phosphatic bed" of Föllmi 1986; Föllmi and Ouwehand 1987; Ouwehand 1987; *pro parte* Plattenwald Bed; on top of the Schrattenkalk Formation), and on the middle inner shelf to the late *tardefurcata* and early *mammillatum* Zones (**Durschlägi Bed**; max. 0.5 m; Fig. 2, section 7 in Fig. 19) and to the *inflatum* Zone (**Wannenalp Bed**; max. 0.3 m; Fig. 2, section 7 in Fig. 19, Fig. 23).
Durschlägi and Wannenalp Beds consist of condensed, winnowed, and of allochthonous particulate phosphatic sediments, including coarse-grained (max. 0.3 mm), glauconitic sandstones and/or micrites (Fig. 23).
In distal parts of the Vorarlberg inner shelf, low sediment accumulation rates persisted throughout most of the Albian, as is documented by the presence of a single, "ultra"-condensed, phosphatic bed (**Plattenwald Bed**; in general: Middle *tardefurcata* to *inflatum* or *dispar* Zones; in distal localities: *jacobi* to *dispar*, *brotzeni*, or *archaeocretacea* Zones, 0.5-1 m, max. 2.3 m; sections 2-6 and 8-13 in Fig. 19; cf. Figs. 7, 15, 20, and 51).
The Plattenwald Bed includes time-equivalent sediments of the Durschlägi, Sellamatt, and Wannnenalp Beds in proximal localities, of the Niederi and Aubrig Beds in addition to these in intermediate localities, and of the Klaus Beds and lower portions of the Seewen Formation in addition to these in distal localities (Fig. 24).

Fig. 21 (next page). Facies changes in the Aubrig Beds toward distal localities. Well-developed Aubrig Beds in Vorarlberg proximal inner shelf localities form a several meters thick, normally graded sequence, followed by a thin, inversely graded sequence, which is transitional to basal beds of the Seewen Formation. Distal Aubrig Beds consist either of both sequences in a more condensed mode (e.g., second and fourth section from the left) or lack basal parts of the normally graded sequence (e.g., middle section). This indicates that distal, seaward portions of the Aubrig sandsheet progressively were subjected to condensation, either by thinning of the whole Aubrig sequence or by including sediments, time-equivalent to the basal Aubrig Beds, in the proceeding Plattenwald condensation. In areas distal to the Aubrig sandsheet, all sediments time-equivalent to Aubrig Beds have been included in the Plattenwald Bed condensation process, as is indicated by the first appearance of the biostratigraphic marker *Rotalipora appenninica* (RENZ) (section to the right)

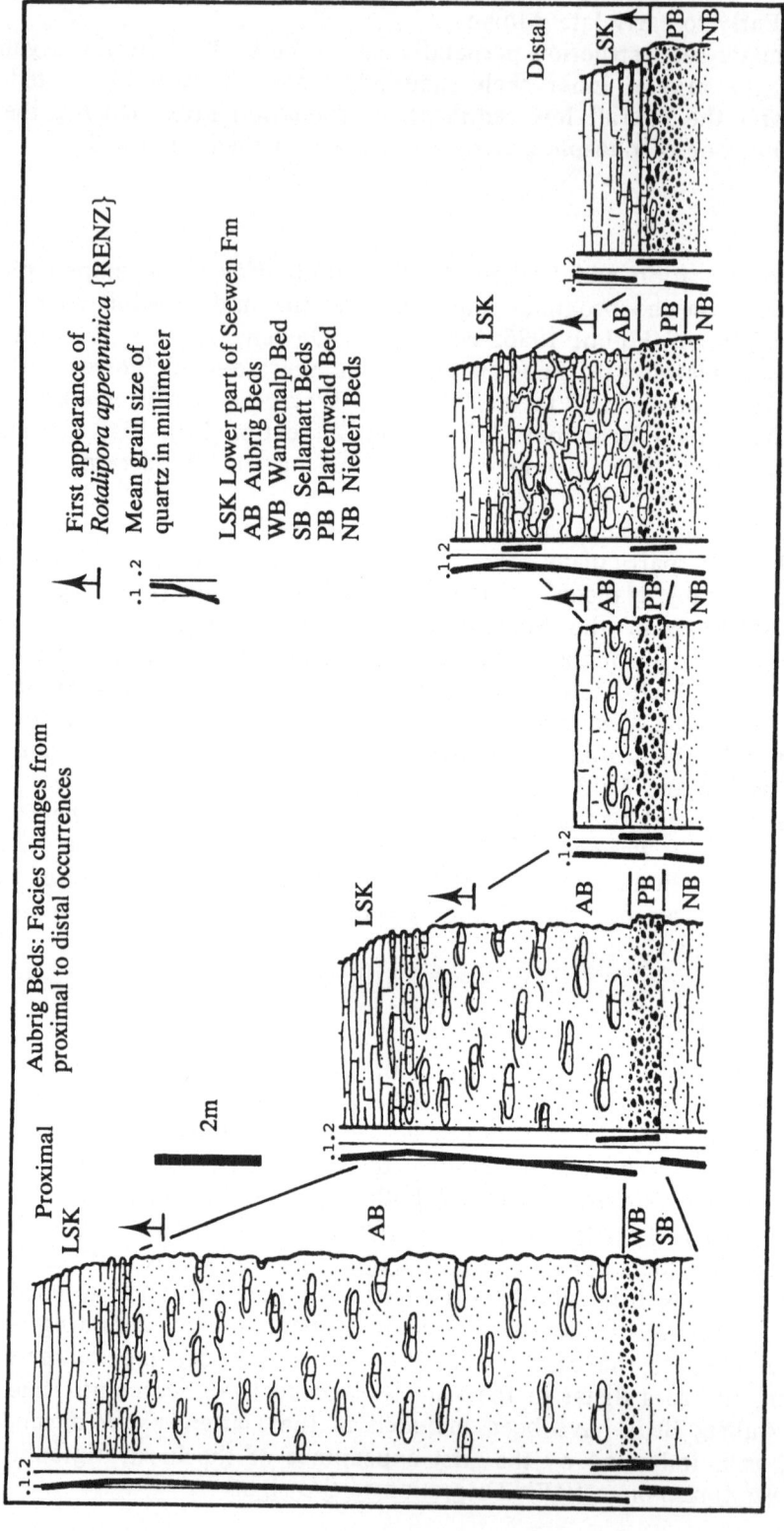

First appearance of
Rotalipora appenninica {RENZ}

.1 .2   Mean grain size of
         quartz in millimeter

LSK   Lower part of Seewen Fm
AB   Aubrig Beds
WB   Wannenalp Bed
SB   Sellamatt Beds
PB   Plattenwald Bed
NB   Niederi Beds

Aubrig Beds: Facies changes from
proximal to distal occurrences

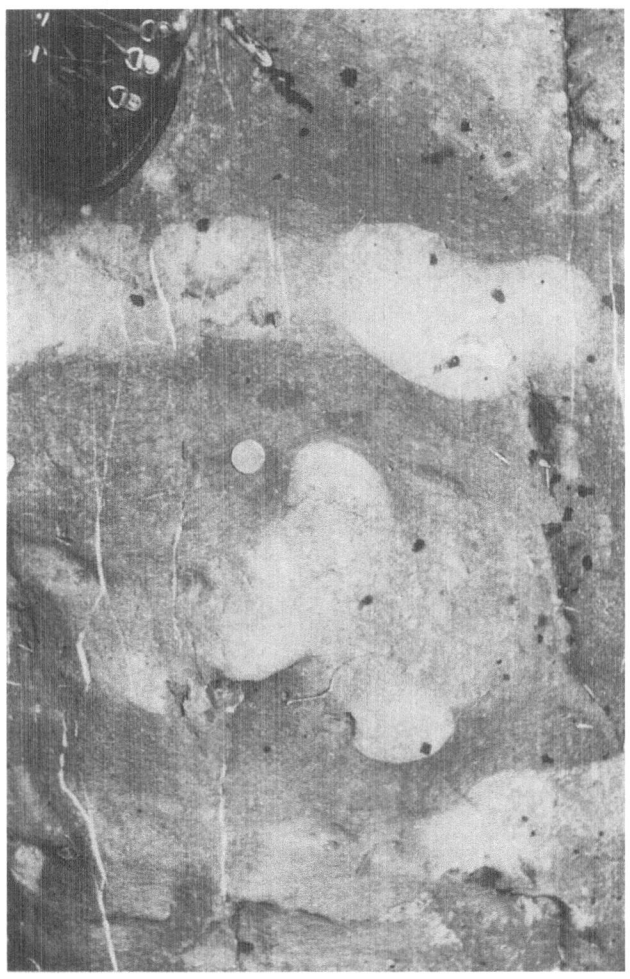

Fig. 22. Aubrig Beds.
Calcareous nodules embedded in a glauconitic sandstone. Parallel to bedding, nodule boundaries are well-defined, due to the presence of stylolites. Perpendicular to bedding, transitional areas are present between the nodules and the surrounding sandstones. The sandstones are depleted in calcareous particles; larger calcareous fossils (e.g., belemnites) show a truncated, irregular periphery (unlike belemnites within the nodules). The uneven distribution of $CaCO_3$, concentrated in layered, pearlstring-like nodular arrangements is due to primary bedding into calcareous richer and poorer horizons (possibly caused by episodic winnowing), and diagenetic overprinting (selective $CaCO_3$ dissolution; coin diameter = 2.5 cm)

Fig. 23. Wannenalp Bed.
Weathered surface of a multi-event winnowed and condensed phosphatic bed.
The phosphatized particles consist of fossil debris (ammonoids: <u>bb</u> = *Beudan-ticeras beudanti* (BRONGNIART); <u>ho</u> = *Hysteroceras* cf. *orbignyi* (SPATH); <u>h</u> = *Hamites* sp.; inoceramids: <u>bc</u> = *Birostrina concentrica* (PARKINSON); <u>bs</u> = *Birostrina sulcata* (PARKINSON); sponges = <u>s</u>; solitary corals = <u>c</u>), which are embedded in a sandy *Hedbergella* micrite

Phosphates of the Plattenwald Bed appear as:

A. Densely packed phosphatized fossil debris and lithoclasts, embedded in nonphosphatized sediments. Commonly, the phosphatic particles display several accreted phosphate generations, which is suggestive of multi-event winnowing (Figs. 24, 25A, and 50). Occasionally, they show signs of abrasion and are mixed with nonphosphatized lithoclasts, which

Fig. 24. Compilation of facies changes within the Plattenwald Bed (<u>left</u> = proximal; <u>right</u> = distal). Lithologic analysis of nonphosphatized sediments included in the Plattenwald Bed (<u>A</u> to <u>H</u>), and indication of their position within the Plattenwald Bed. Note the following trends toward distal localities: (1) the youngest phosphatized ammonoids progressively become older; (2) the diversity of nonphosphatized Plattenwald sediments decreases; and (3) detritus-rich sediments disappear, and calcareous sediments dominate

Plattenwald Bed: Facies changes from proximal to distal occurrences

Proximal inner shelf ←   → Distal inner shelf

KB Klaus Beds
BB Brisi Beds
SKF Schrattenkalk Fm
LSK Lower part of Seewen Fm
AB Aubrig Beds
NB Niederi Beds

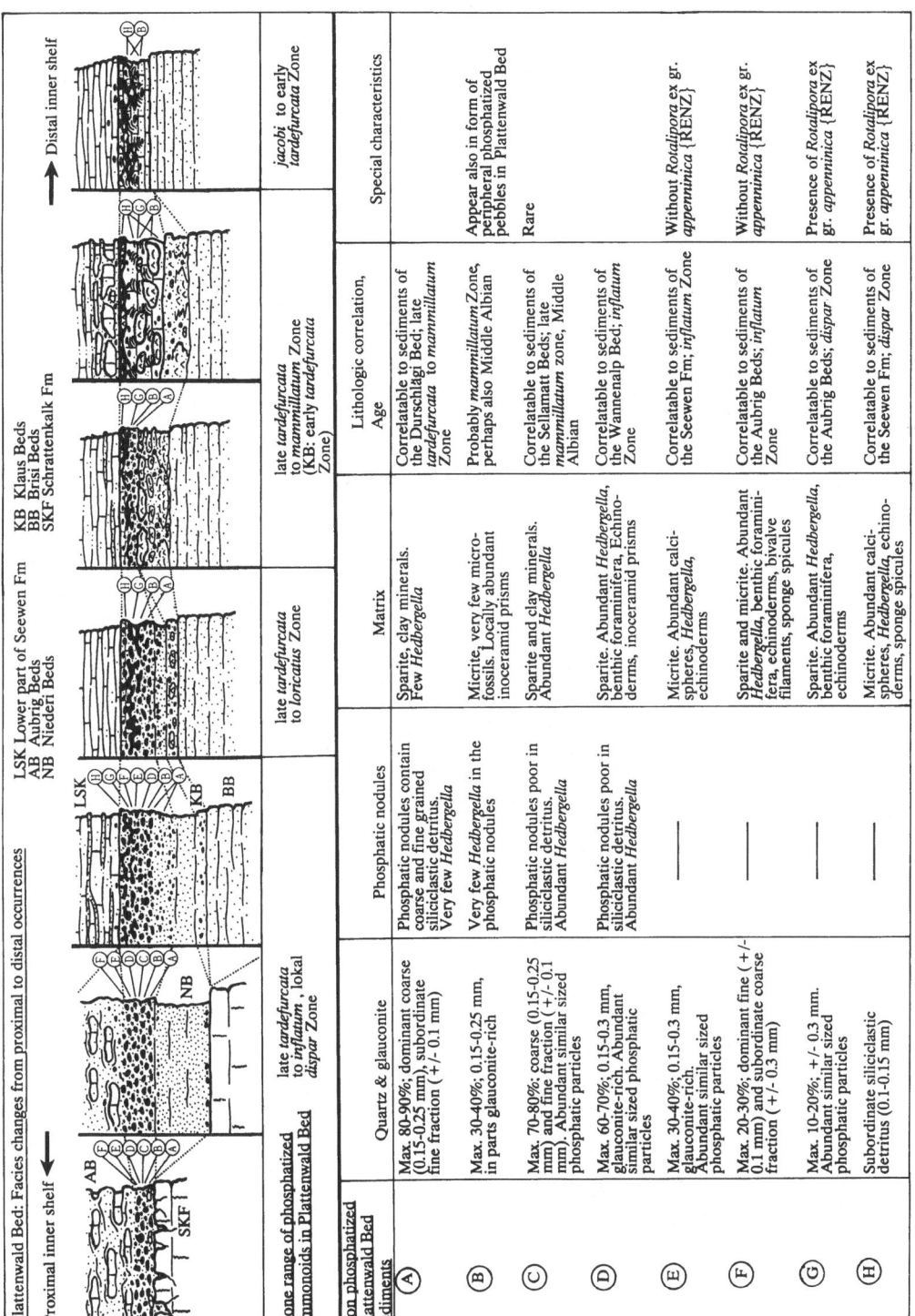

Zone range of phosphatized ammonoids in Plattenwald Bed	late *tardefurcata*, lokal to *inflatum*, *dispar* Zone	late *tardefurcata* to *loricatus* Zone	late *tardefurcata* Zone	late *tardefurcata* to *mammillatum* Zone (KB: early *tardefurcata* Zone)	*jacobi* to early *tardefurcata* Zone

Non phosphatized Plattenwald Bed sediments	Quartz & glauconite	Phosphatic nodules	Matrix	Lithologic correlation, Age	Special characteristics
A	Max. 80-90%; dominant coarse (0.15-0.25 mm), subordinate fine fraction (+/- 0.1 mm)	Phosphatic nodules contain coarse and fine grained siliciclastic detritus. Very few *Hedbergella*	Sparite, clay minerals. Few *Hedbergella*	Correlatable to sediments of the Durschlägi Bed; late *tardefurcata* to *mammillatum* Zone	
B	Max. 30-40%; 0.15-0.25 mm, in parts glauconite-rich	Very few *Hedbergella* in the phosphatic nodules	Micrite, very very few micro-fossils. Locally abundant inoceramid prisms	Probably *mammillatum* Zone, perhaps also Middle Albian	Appear also in form of peripheral phosphatized pebbles in Plattenwald Bed
C	Max. 70-80%; coarse (0.15-0.25 mm) and fine fraction (+/- 0.1 mm). Abundant similar sized phosphatic particles	Phosphatic nodules poor in siliciclastic detritus. Abundant *Hedbergella*	Sparite and clay minerals. Abundant *Hedbergella*	Correlatable to sediments of the Sellamatt Beds; late *mammillatum* zone, Middle Albian	Rare
D	Max. 60-70%; 0.15-0.3 mm, glauconite-rich. Abundant similar sized phosphatic particles	Phosphatic nodules poor in siliciclastic detritus. Abundant *Hedbergella*	Sparite. Abundant *Hedbergella*, benthic foraminifera, Echinoderms, inoceramid prisms	Correlatable to sediments of the Wannenalp Bed; *inflatum* Zone	
E	Max. 30-40%; 0.15-0.3 mm, glauconite-rich. Abundant similar sized phosphatic particles	—	Micrite. Abundant calcispheres, *Hedbergella*, echinoderms	Correlatable to sediments of the Seewen Fm; *inflatum* Zone	Without *Rotalipora* ex gr. *appenninica* {RENZ}
F	Max. 20-30%; dominant fine (+/- 0.1 mm) and subordinate coarse fraction (+/- 0.3 mm)	—	Sparite and micrite. Abundant *Hedbergella*, benthic foraminifera, echinoderms, bivalve filaments, sponge spicules	Correlatable to sediments of the Aubrig Beds; *inflatum* Zone	Without *Rotalipora* ex gr. *appenninica* {RENZ}
G	Max. 10-20%; +/- 0.3 mm. Abundant similar sized phosphatic particles	—	Sparite. Abundant *Hedbergella*, benthic foraminifera, echinoderms	Correlatable to sediments of the Aubrig Beds; *dispar* Zone	Presence of *Rotalipora* ex gr. *appenninica* {RENZ}
H	Subordinate siliciclastic detritus (0.1-0.15 mm)	—	Micrite. Abundant calcispheres, *Hedbergella*, echinoderms, sponge spicules	Correlatable to sediments of the Seewen Fm; *dispar* Zone	Presence of *Rotalipora* ex gr. *appenninica* {RENZ}

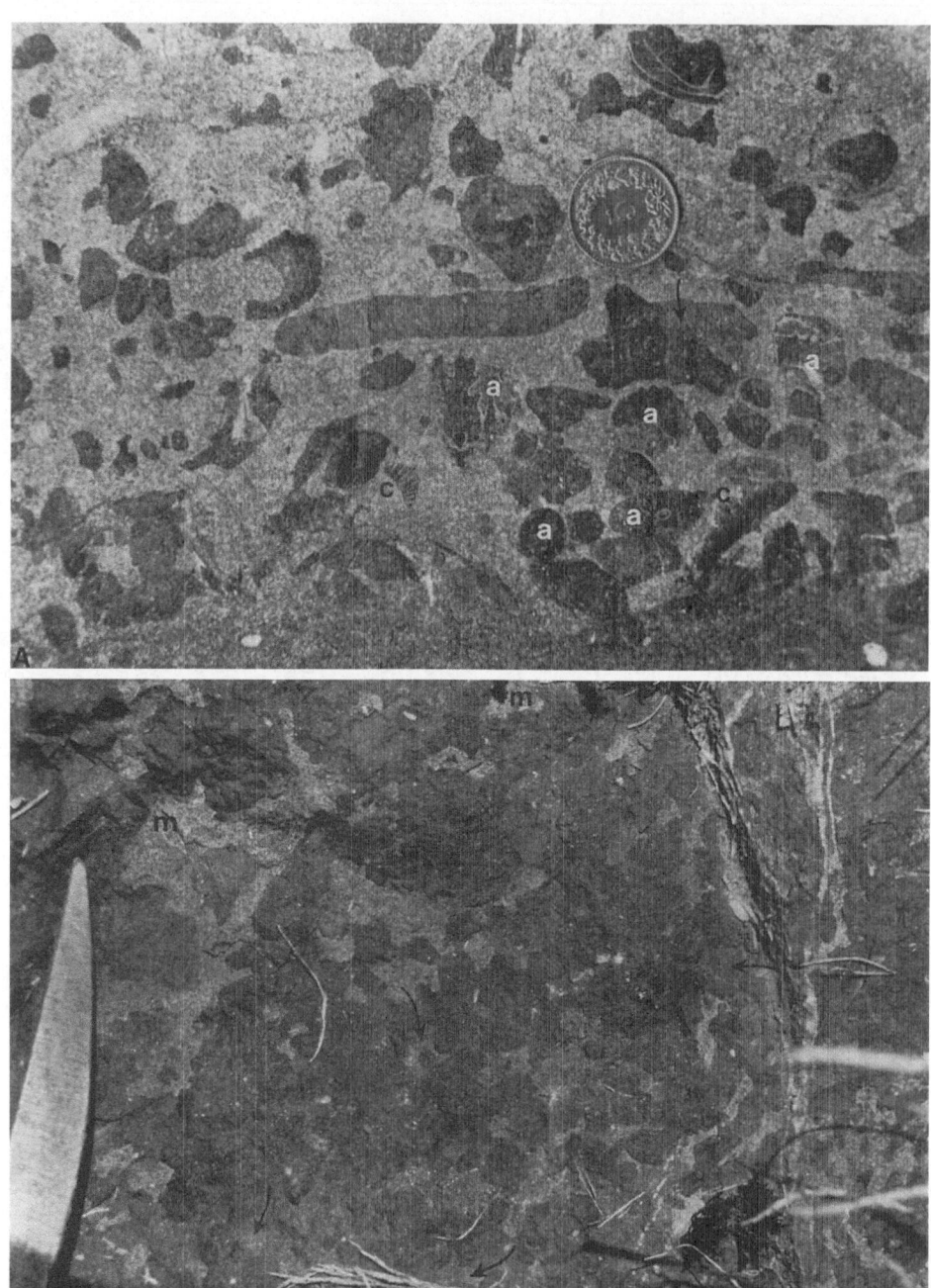

indicates an allochthonous character for these particles (Föllmi et al. in press).

B. A thin (<10 cm) phosphatic crust, which consists of phosphatized microbial mats, encrusting and/or sessile organisms, and phosphatized, trapped sediments (Figs. 26, 27, and 52).

Transitions between or mixtures of these end members are very common: A can be superposed by B and vice versa; A can subsequently be cemented by phosphatized sediments (Fig. 25B); B can be disrupted and transformed into A, amongst other variations (Figs. 19, 24, and 50). However, winnowed and transported nodular sediments are generally dominant in western parts of the Vorarlberg distal inner shelf, whereas autochthonous phosphatic crusts are well developed in Plattenwald Bed localities of the eastern Vorarlberg distal inner shelf. This is suggestive of eastwardly increasing hydraulic energy levels during deposition of the Plattenwald Bed, which is also indicated by the eastward disappearance of the contemporaneous Sellamatt Beds (see above; Sect. 3.4).

Two trends are observed in the distribution of nonphosphatized sediments within the phosphatic Plattenwald Bed: The diversity of nonphosphatized sediments increases with proximality of the Plattenwald Bed, and detritus-rich sediments prevail in proximal settings, whereas micritic sediments dominate in distal settings (Fig. 24).

The age of nonphosphatized sediments is not necessarily coincident with that of the included phosphatic particles. In general, these sediments are younger; in distal localities they are distinctly younger (Figs. 24 and 28). Even nonphosphatized sediments, subjacent to phosphatic crusts, are in some cases younger. This age "inversion" is probably due to the effect of "tunneling" hydraulic erosive forces, induced by scouring currents that removed less consolidated sediments underneath the phosphatic crust (Fig. 28).

The time span in which phosphogenesis occurred (= T_{PH}) is estimated by

Fig. 25. Plattenwald Bed.
A. Phosphatic particles embedded in a glauconitic sandstone. They generally consist of fossil debris (a ammonoids; c solitary corals; other unidentifiable fossil debris). Some nodules display more than one phosphate generation (arrow). Ammonoids in this locality indicate the period between *tardefurcata* and *inflatum* Zones. The concentration of phosphate particles is due to episodic winnowing (surface approximately parallel to bedding; coin diameter = 1.5 cm).
B. Winnowed phosphatic particles embedded in glauconitic sands and cemented by different phosphate generations (arrows). Included phosphatized ammonoids indicate the time span between *tardefurcata* and *mammillatum* Zones. Depressions are infilled with a younger, nonphosphatized sediment (m sandy, glauconitic micrite; surface approximately parallel to bedding)

Fig. 26. Phosphatized microbial mat in the Plattenwald Bed.
A. Phosphatic crust (c) overgrown by phosphatized columnar microbial mats.
B. Close up

the zone range of phosphatized ammonoids and inoceramid bivalves present in the Plattenwald Bed. Comparison of this time span with the overall duration of condensation (= T_C), estimated by globotruncanid foraminifera within the nonphosphatized sediments, reveals that

A. In proximal Plattenwald Bed localities, T_{PH} = T_C = late *tardefurcata* to *inflatum*, local *dispar* Zone (zone ages in Fig. 2; area D in Fig. 19).
B. In intermediate localities of western Vorarlberg, T_{PH} = late *tardefurcata* to *loricatus* Zone, whereas T_C = late *tardefurcata* to *dispar* Zone (area C in Fig. 19).
C. In distal localities, T_{PH} = late *tardefurcata* to *mammillatum* Zone, and T_C = late *tardefurcata* to *dispar*, *brotzeni*, and locally to *archaeocretacea* Zone. To the southeast of this area, the Plattenwald Bed includes reworked Klaus Beds (Fig. 28), which resets the beginning of T_{PH} and T_C to the *jacobi* Zone (area B in Fig. 19).
D. In the interchannel areas along the Rankweil Ramp, T_{PH} = *jacobi* to early *tardefurcata* Zone, whereas T_C = *jacobi* to *dispar*, and locally to *archaeocretacea* Zone (area A in Fig. 19). Unfortunately, most Plattenwald sediments in this zone were obliterated during subsequent erosion events (questionmarks in Fig. 19).

Phosphate deposition in the Plattenwald Bed appears to have depended on the presence of terrigenous, siliciclastic detritus (Fig. 24). Micritic, non-phosphatized sediments, included in the Plattenwald Bed, are generally younger than the youngest phosphatic particles. Phosphatic particles in distal Plattenwald Bed sections, which lack nonphosphatized siliciclastic sediments still include siliciclastic detritus, which is suggestive of the temporary presence of siliciclastic fractions in this part of the inner shelf (during the time interval of phosphogenesis). Subsequently, the sands may have been removed by current activity and incorporated into the distal Rankweil Beds (see below; Sects. 3.4 and 3.6.2).

3. Channels, incised into the Rankweil Ramp during the latest Aptian, maintained stable positions and were actively infilled during most of the Albian. Channel infills (uppermost Aptian to uppermost Albian Rankweil Beds) include coarse to fine-grained glauconitic sands, phosphatic particles, reworked granules and pebbles of older sediments (e.g., Gams Beds, Schrattenkalk, and Drusberg Formations), nonphosphatized fossil debris, and locally micrite-rich zones. The Rankweil Beds encompass gravity flow deposits, derived as "spill-over" sediments, bypassing the inner shelf (e.g., Hamilton et al. 1980; Flemming 1981). The channels also served as a trap for pelagic sediments (sections 14 and 15 in Fig. 19).

4. In the outer shelf, sedimentation of dark, commonly laminated and poorly bioturbated marls and muds continued (upper portion of the Freschen Beds; Lower to Upper Albian; sections 16 and 17 in Fig. 19). In contrast to the middle late Aptian, deposition of the Freschen Beds extended into the distal outer shelf (documented in the ultrahelvetic "Hohe Kugel" area), where they cover Hochkugel Beds (section 17 in Fig. 19).

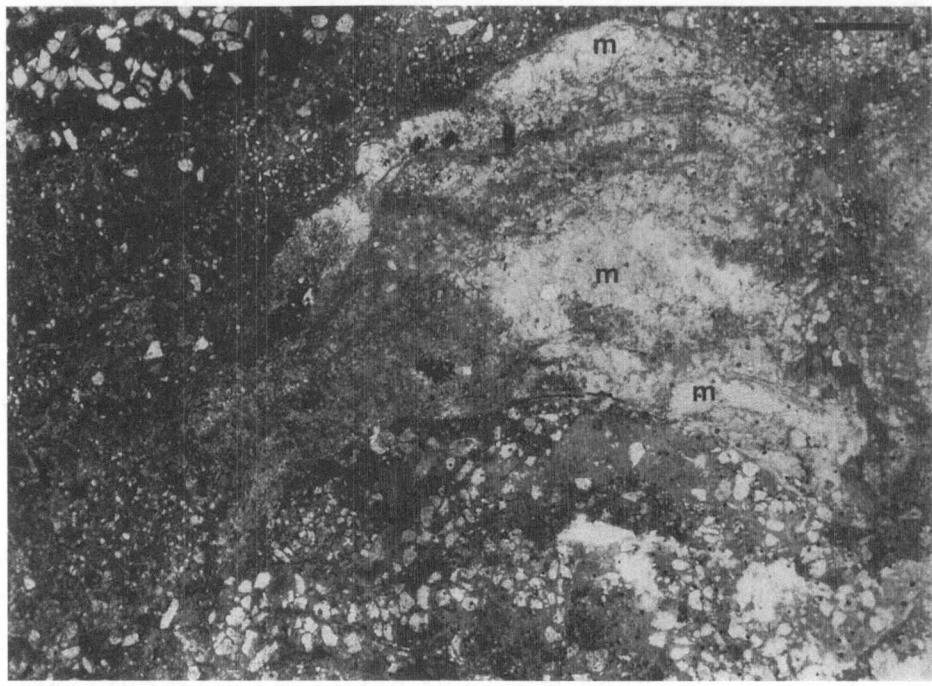

Fig. 27. Phosphatized microbial colony in Plattenwald Bed.
Thin-section photomicrograph (bar = 1 mm): Colonization of a phosphatic particle consisting of several phosphate generations, by a subsequently phosphatized microbial assemblage (m)

The Albian sediment accumulation rates of the Freschen Beds are significantly lower than the middle Upper Aptian to lowermost Albian ones (Albian: 1.4-1.6 m/my; middle Upper Aptian to lowermost Albian: 5-10 m/my). This is related to a generally lower influx of terrigenous detritus during the Albian, which is possibly due to elevated sea levels (Sect. 3.7). Proximal Albian Freschen Beds interfinger with distal Rankweil Beds (e.g., sections 13 and 14 in Fig. 7).

Fig. 28. Sequential diagram of the depositional evolution of the Klaus and Plattenwald Beds in distal, southeastern localities of the Vorarlberg Plattenwald Bed distribution area (Fig. 19). Note that age inversions are induced by the effect of "tunneling" erosion

53

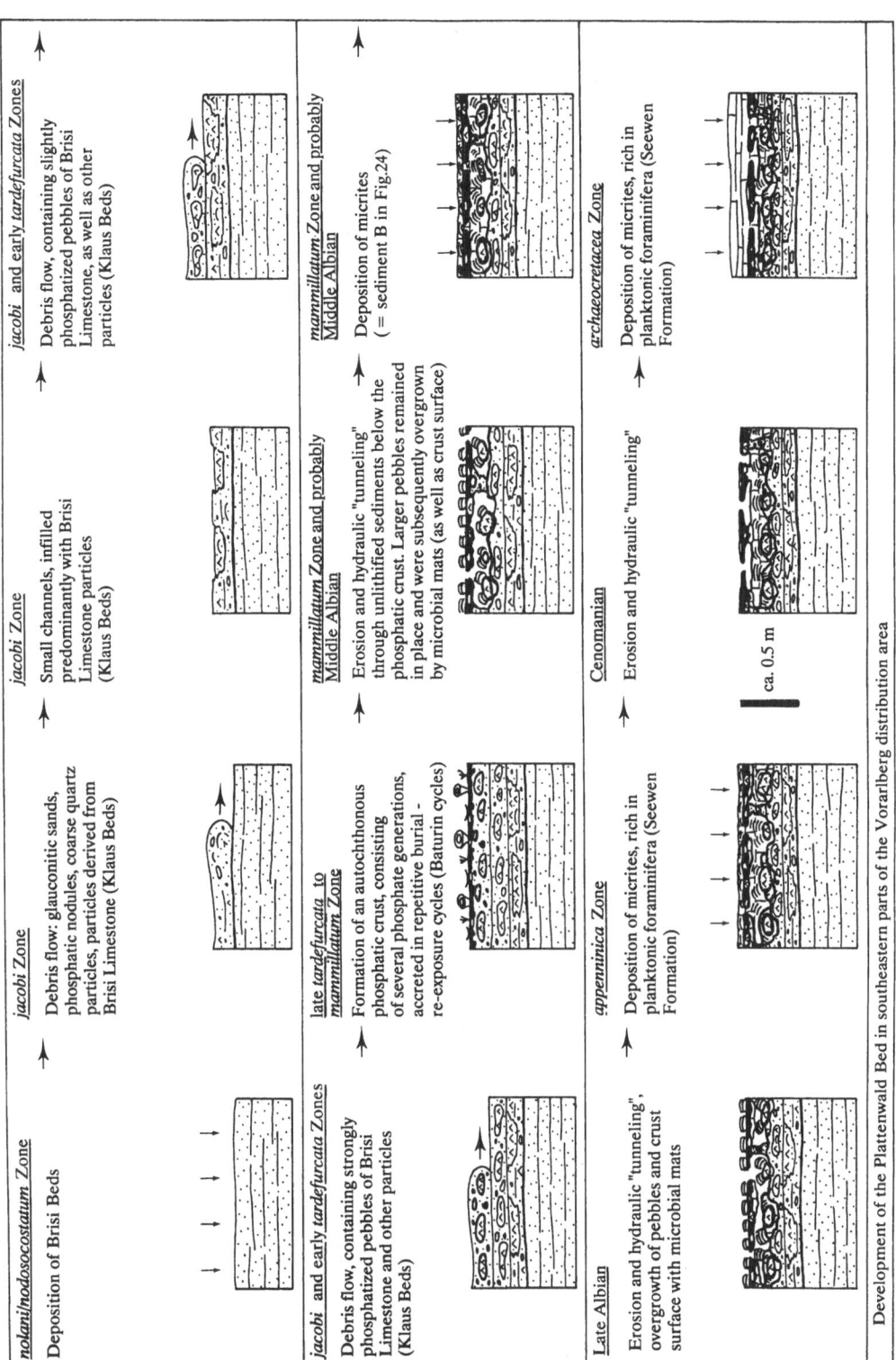

Development of the Plattenwald Bed in southeastern parts of the Vorarlberg distribution area

nola/i/nodosocostatum Zone

Deposition of Brisi Beds

jacobi Zone

Debris flow: glauconitic sands, phosphatic nodules, coarse quartz particles, particles derived from Brisi Limestone (Klaus Beds)

jacobi Zone

Small channels, infilled predominantly with Brisi Limestone particles (Klaus Beds)

jacobi and early *tardefurcata* Zones

Debris flow, containing slightly phosphatized pebbles of Brisi Limestone, as well as other particles (Klaus Beds)

jacobi and early *tardefurcata* Zones

Debris flow, containing strongly phosphatized pebbles of Brisi Limestone and other particles (Klaus Beds)

late *tardefurcata* to *mammillatum* Zone

Formation of an autochthonous phosphatic crust, consisting of several phosphate generations, accreted in repetitive burial - re-exposure cycles (Baturin cycles)

mammillatum Zone and probably Middle Albian

Erosion and hydraulic "tunneling" through unlithified sediments below the phosphatic crust. Larger pebbles remained in place and were subsequently overgrown by microbial mats (as well as crust surface)

mammillatum Zone and probably Middle Albian

Deposition of micrites (= sediment B in Fig.24)

Late Albian

Erosion and hydraulic "tunneling", overgrowth of pebbles and crust surface with microbial mats

appenninica Zone

Deposition of micrites, rich in planktonic foraminifera (Seewen Formation)

Cenomanian

Erosion and hydraulic "tunneling"

ca. 0.5 m

archaeocretacea Zone

Deposition of micrites, rich in planktonic foraminifera (Seewen Formation)

2.6.2 Time-Space Relations

The succession of condensed phosphatic beds (inner shelf), proximally in-
terlayered with glauconitic sandstones and marls, channelized redeposited sediments (along the Rankweil Ramp and on the proximal outer shelf), and
commonly laminated, dark muds and marls (outer shelf) constitutes a classical
current-dominated depositional system, comparable to that of the middle early
to early late Aptian (Figs. 20, 49, and 51). This system was established in the
late *tardefurcata* Zone, and persisted throughout the Albian (until the *dispar*
Zone; approximately 13 my after Harland et al. 1982), as is indicated by the
following evidence:

1. Ammonoids, preserved in the Plattenwald Bed, indicate each Albian ammonite zone (late *tardefurcata* to *dispar* Zone: e.g., *Leymeriella, Douvillei-
   ceras, Otohoplites, Hoplites, Anahoplites, Euhoplites, Brancoceras, Mortoniceras, Hysteroceras, Oxytropidoceras, Stoliczkaia*; Föllmi 1986, 1989). Distal
   Plattenwald Bed sections include older ammonoids, which are probably reworked from the Klaus Beds (e.g., *Hypacanthoplites*).
2. Rankweil Beds contain middle and early late Albian ammonoids and inoceramids such as *Birostrina concentrica* and *B. sulcata* (PARKINSON), Locally, Rankweil Beds also include latest Aptian and earliest Albian ammonoids in a basal conglomeratic layer (Sect. 2.5.2).
3. The upper sequence of the Freschen Beds (Lower to Upper Albian) is characterized by the abundance of inoceramid debris and prisms, whereas the
   lower sequence of the Freschen Beds (middle Upper Aptian to lowermost
   Albian) lack such remains. The first appearance of inoceramid bivalves on
   the helvetic shelf date back to the late *tardefurcata* Zone, where they occupied the ecological niche of the Aptian to earliest Albian genus *Aucellina*
   (Kemper 1982; Föllmi 1986).
4. Nonphosphatized, micritic sediments within the Plattenwald Bed (sediments
   G and H in Fig. 24) and on top of the Rankweil and Freschen Beds include
   the globotruncanid foraminifera *Rotalipora* ex gr. *praeticinensis - ticinensis*
   (GANDOLFI) and *R. appenninica* (RENZ), indicative of the *appenninica* .
   Zone (approximately time-equivalent to the *dispar* Zone; Fig. 2).

Fig. 29. Latest Albian to early Cenomanian.
Palinspastic map displays the distribution of the Kamm Bed and Seewen Formation during the Albian-Cenomanian transition. The Rankweil Ramp and the
proximal outer shelf area are characterized by zero net sediment accumulation
rates and subordinate erosion during this period and most of the Cenomanian
(section 4). The sections in this figure show typical transitions from the Garschella into the Seewen Formation, as well as typical sequences in the lower
part of the Seewen Formation (Cenomanian; exceptions are sections 1 and 4).
Correlation of the uppermost Albian to Lower Cenomanian sediments is indicated by dots

55

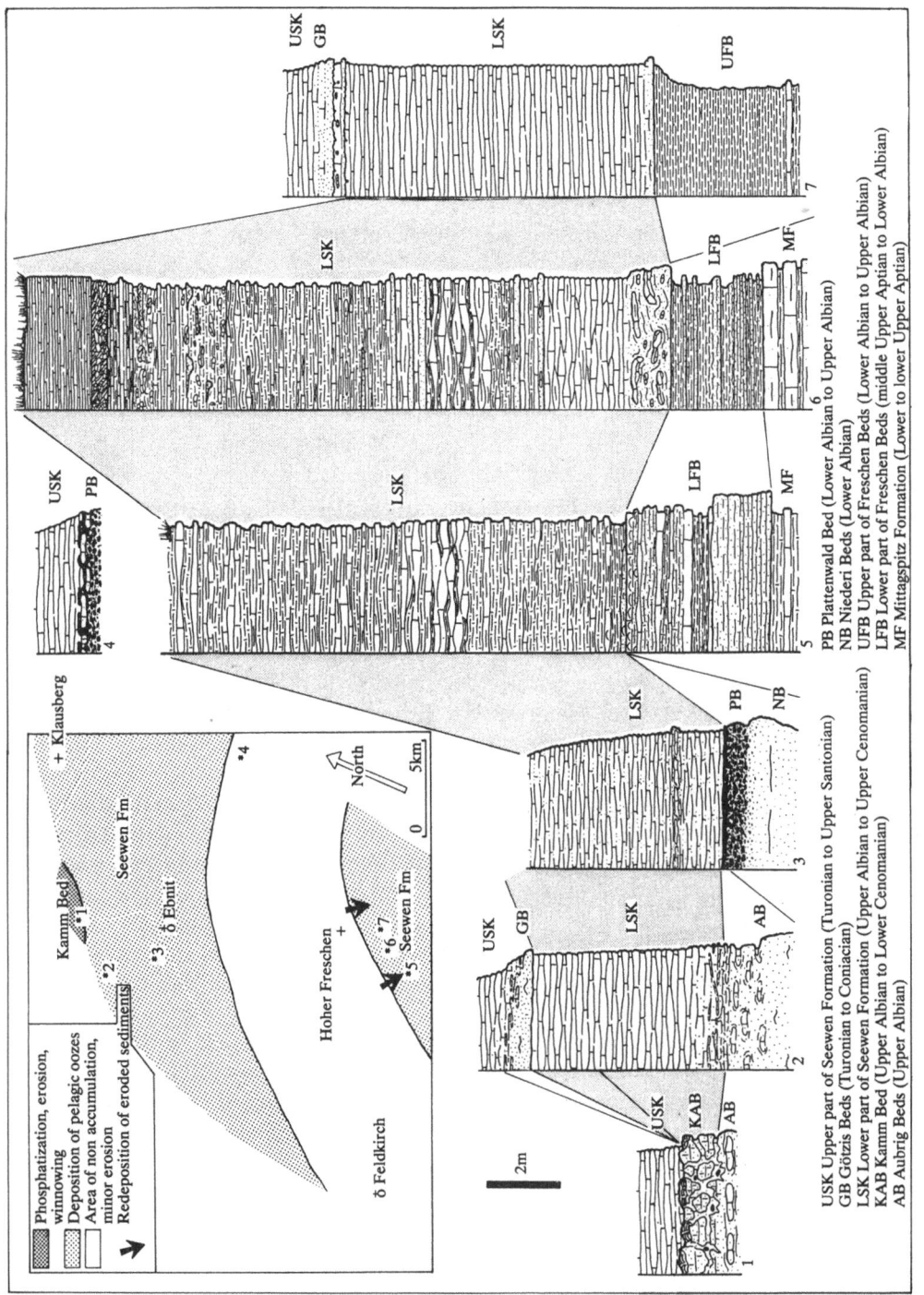

USK Upper part of Seewen Formation (Turonian to Upper Santonian)
GB Götzis Beds (Turonian to Coniacian)
LSK Lower part of Seewen Formation (Upper Albian to Upper Cenomanian)
KAB Kamm Bed (Upper Albian to Lower Cenomanian)
AB Aubrig Beds (Upper Albian)

PB Plattenwald Bed (Lower Albian to Upper Albian)
NB Niederi Beds (Lower Albian)
UFB Upper part of Freschen Beds (Lower Albian to Upper Albian)
LFB Lower part of Freschen Beds (middle Upper Aptian to Lower Albian)
MF Mittagspitz Formation (Lower to lower Upper Aptian)

2.7 Latest Albian to Early Cenomanian

At the Albian-Cenomanian boundary, the helvetic shelf experienced a gradual, diachronous alteration in sediment patterns with the change from the formation of condensed glauconitic and phosphatic beds, redeposited sediments, and laminated dark shales (= Garschella Formation), to the accumulation of predominantly pelagic, micritic sediments (= Seewen Formation). This remarkable change started in the latest Albian with a prominent shift of the "phosphatization-erosion-winnowing-redistribution" sedimentary system toward the north, toward proximal areas of the inner shelf, distal inner shelf areas subsequently being occupied by the pelagic regime (Figs. 29, 30, and 49).

1. In proximal inner shelf areas, a fossiliferous, condensed and winnowed, nodular, phosphatic bed developed on top of the Aubrig Beds (limited to eastern Swiss outcrops; uppermost Albian to Lower Cenomanian **Kamm Bed**; max. 0.5 m; Figs. 2, 29, and 30; Föllmi and Ouwehand 1987; Ouwehand 1987). In a distal belt of their distribution area, Kamm Bed sediments include abundant and widely distributed, fossilized and well-preserved, columnar microbial mats, which are attached to the rugged and phosphatized surface of the subjacent Aubrig Beds. The mats are preserved in slightly phosphatized, pelagic, Seewen-like micrites (section 1 in Fig. 29; Fig. 31; Ouwehand 1987). This facies extends into the northernmost part of the Vorarlberg helvetic inner shelf (Figs. 29 and 30).
2. In intermediate and distal parts of the inner shelf, deposition of a pelagic calcareous ooze started (**Seewen Formation**; lower sequence = uppermost Albian to uppermost Cenomanian; max. 20 m; Fig. 2; sections 2, 3 in Fig. 29; Fig. 30).
   In the intermediate inner shelf, south of the Kamm Bed distribution area, the oldest Seewen sediments, on top of the Aubrig Beds, date from the *brotzeni* Zone (early Cenomanian) and sediments with an *appenninica* Zone age are confined to the uppermost Aubrig Beds and to a thin transitional sequence between the Aubrig Beds and the Seewen Formation.
   In distal inner shelf areas, the oldest Seewen sediments, on top of the Plattenwald Bed, commonly display an uppermost Albian age (*appenninica* Zone).
   Proximal basal Seewen sediments and transitional sediments to the subjacent Aubrig Beds (*appenninica* and *brotzeni* Zones) are thinner compared to their distal inner shelf, time-equivalent counterparts. This is probably due to mild condensation, related to the formation of the adjacent Kamm Bed (Fig. 29).

Fig. 30. Latest Albian to early Cenomanian.
Diagrammatic reconstruction perpendicular to the shelf. Note the landward shift in the area of phosphogenesis compared to the Albian (Fig. 20), and the onlap of pelagic micrites onto the inner shelf

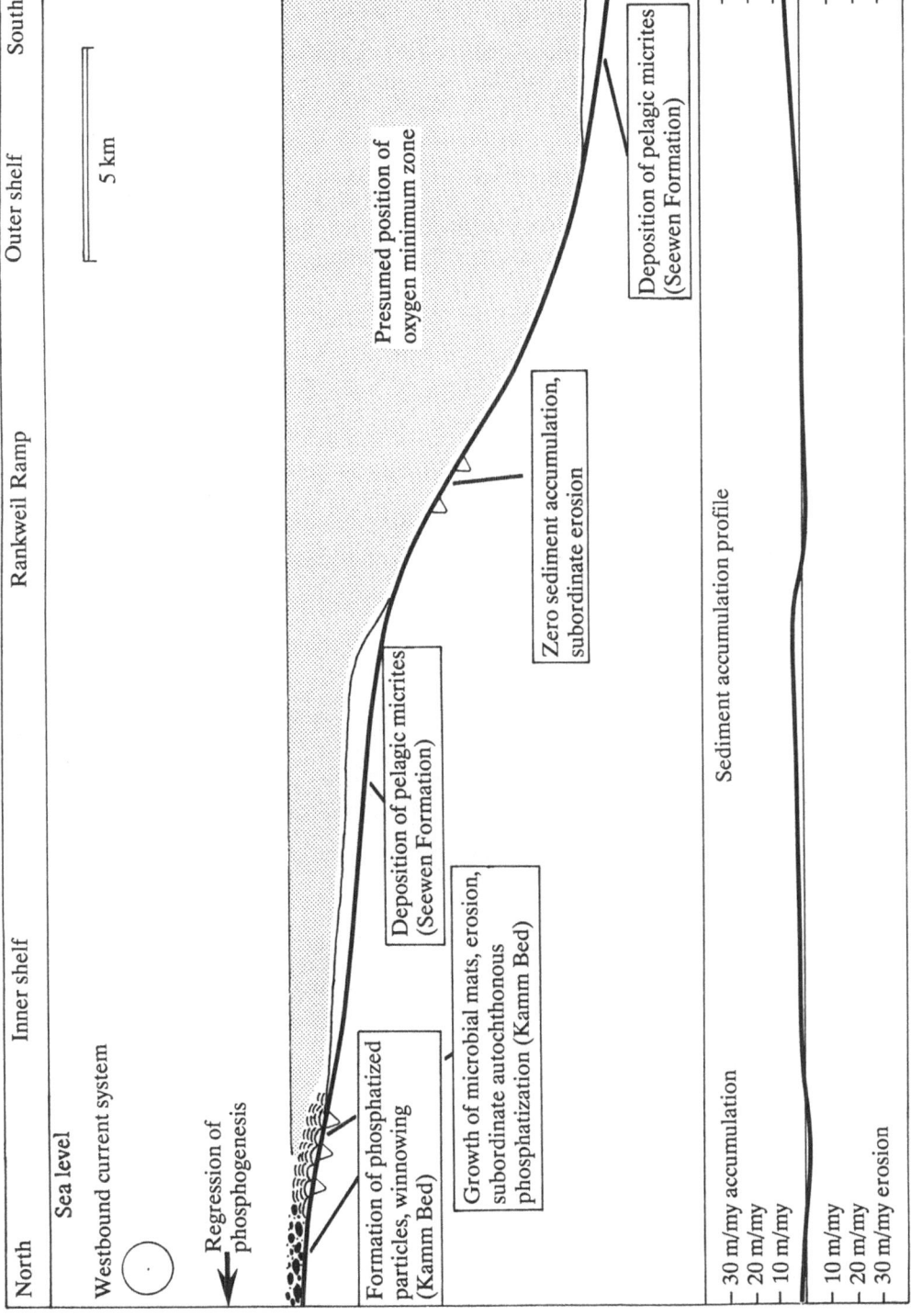

3. The channels perpendicular to the Rankweil Ramp were at this time infilled and inactivated. The sediment accumulation rates along the ramp approached zero; local erosion was common during this period and during most of the following Cenomanian (see below; section 4 in Fig. 29; Fig. 30).
4. During the Albian–Cenomanian transition, Seewen sediments blanketed the entire outer shelf, the oldest being coeval to their inner shelf counterparts (*appenninica* Zone; Figs. 29 and 30). Proximal outer shelf Seewen sediments from this period are typically rich in clays and include channelized and sheet-like gravity flow deposits (max. 20 cm thick). These event deposits consist of coarse-grained (max. 0.6 mm), glauconitic sands and phosphatic particles, which most probably were derived from exposed Rankweil Beds along the adjacent Rankweil Ramp (e.g., sections 5 and 6 in Fig. 29).

2.8 Middle to Late Cenomanian

The deposition of pelagic Seewen sediments continued in the *reicheli* and *cushmani* Zones and expanded onto proximal areas of the inner shelf, thus terminating the youngest phase of phosphogenesis and condensation, associated with the Garschella Formation (youngest Kamm Bed sediments date from the *reicheli* Zone; Föllmi and Ouwehand 1987). By the end of the *reicheli* Zone,

Fig. 31. Kamm Bed.
Polished Kamm Bed specimen. *Thalassinoides*-like fissures in uppermost Aubrig Beds infilled with Seewen-type micrites. Irregularly colored structures and laminations within micrites represent preserved microbial mats. The Aubrig sediments are surficially phosphatized

the pelagic regime is established across the entire helvetic shelf. Along the Rankweil Ramp, sediment accumulation rates remained zero (Figs. 2, 29, and 30).

2.9 Latest Cenomanian to Earliest Turonian

The latest Cenomanian to earliest Turonian interval (*archaeocretacea* Zone) is recognized as a period of widespread anoxia, characterized by a major carbon-isotope shift (e.g., Scholle and Arthur 1980; De Graciansky et al. 1981, 1986, 1987; Arthur et al. 1987, 1988; Schlanger et al. 1987; Bralower 1988). Although this shift was detected recently in the Seewen Formation (Greber 1987), true anoxic sediments are not known from the helvetic shelf. Instead, the *archaeo-cretacea* Zone represents an episode of erosion and redistribution of eroded sediments, incorporated in a sandstone bed of surprisingly uniform appearance throughout the Vorarlberg helvetic shelf (*archaeocretacea* Zone **Götzis Bed**; max. 2 m; Figs. 2, 32, and 33). The Götzis Bed consists of a well-sorted, non- to normally graded (max. 0.6 mm), glauconitic sandstone. A distinct truncational unconformity separates the Götzis Bed from subjacent Seewen sediments (of *cushmani* or *reicheli* Zone age; Figs. 32 and 33). Syndepositional *Thalassinoides* burrows, infilled with glauconitic sand, pipe down into the underlying Seewen sediments (Fig. 33B). The unconformity embraces a hiatus, which extends from the *reicheli* or *cushmani* Zone to the *archaeocretacea* Zone. Seewen sediments above the Götzis Bed include autochthonous planktonic foraminifera from the *helvetica* Zone, or reworked foraminifera from the *archaeocretacea* and/or *cushmani* Zone.

The occurrence of the *archaeocretacea* Zone Götzis Bed is limited to the western part of the Vorarlberg inner shelf and to the outer shelf. The Rankweil Ramp, and central and eastern parts of the Vorarlberg inner shelf were affected by younger erosive episodes, in which sediments, time-equivalent to the *archaeocretacea* Zone Götzis Bed, were radically obliterated (Figs. 32 and 34). Proximal inner shelf outcrops in eastern Switzerland (Blumer 1905) suggest that the *archaeocretacea* Zone Götzis Bed extends over at least 20 km in a landward direction, but disappears in a westward direction, parallel to the Rankweil Ramp, within 5-8 km (Föllmi 1981, 1986).

Fig. 32 (next page). Latest Cenomanian to early Turonian.
In the palinspastic map, the distribution areas of the *archaeocretacea* Zone Götzis Bed are shown. The areas are limited to the western part of the inner shelf and to the outer shelf. The area in between (including the Rankweil Ramp) was affected by younger, erosive episodes (especially by the late Turonian-Coniacian erosive event), in which sediments of the Cenomanian-Turonian boundary interval were obliterated. The map displays the age of Seewen sediments immediately subjacent to the Götzis Bed (*reicheli* and *cushmani* Zones), indicating that amounts of eroded sediment increase toward the central inner shelf and the Rankweil Ramp (Fig. 34)

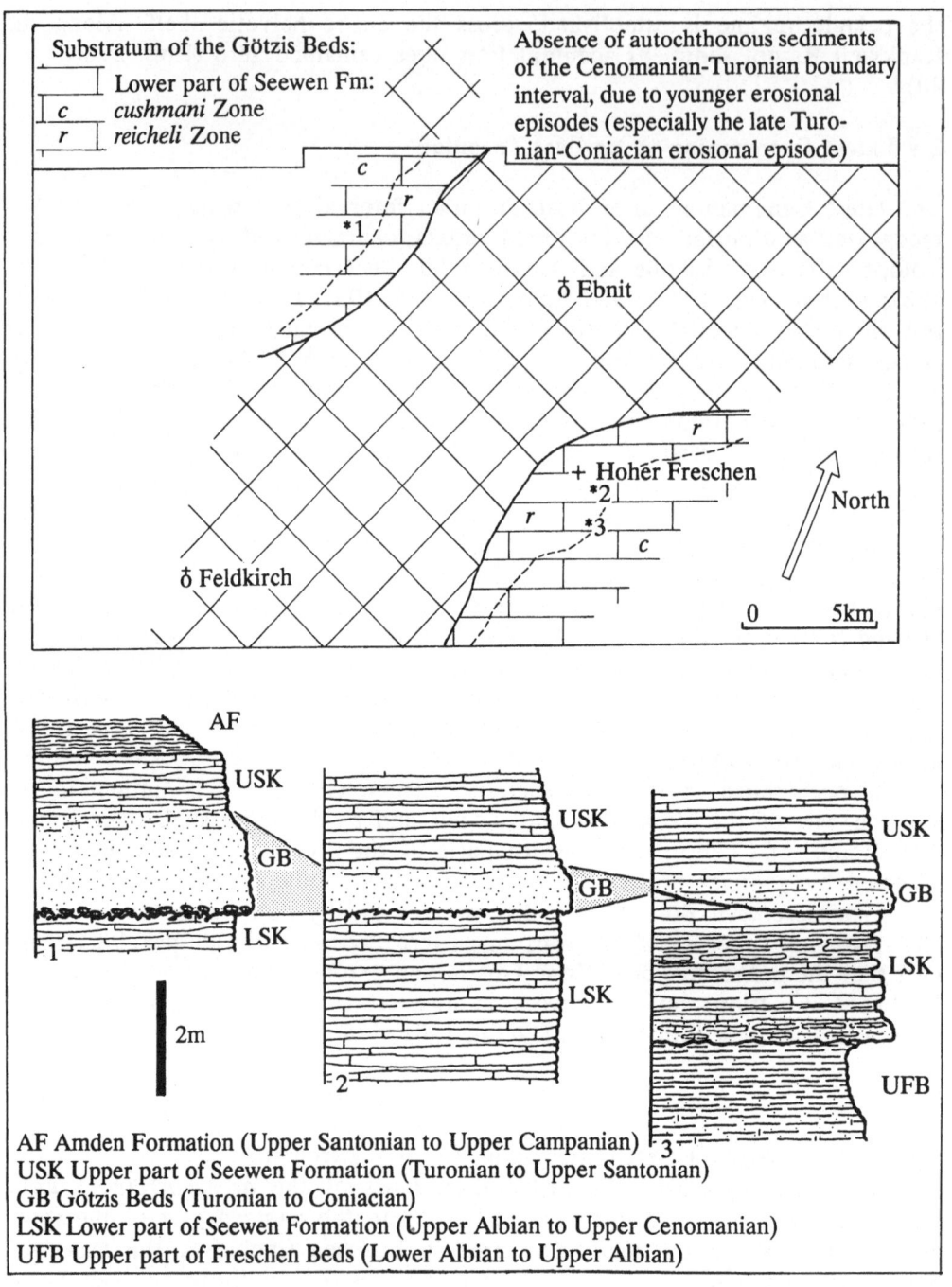

Substratum of the Götzis Beds:

Lower part of Seewen Fm:
c *cushmani* Zone
r *reicheli* Zone

Absence of autochthonous sediments of the Cenomanian-Turonian boundary interval, due to younger erosional episodes (especially the late Turonian-Coniacian erosional episode)

ð Ebnit

+ Hoher Freschen
*2
r *3
c

ð Feldkirch

North

0   5km

AF
USK
GB
LSK
2m
1

USK
GB
LSK
2

USK
GB
LSK
UFB
3

AF Amden Formation (Upper Santonian to Upper Campanian)
USK Upper part of Seewen Formation (Turonian to Upper Santonian)
GB Götzis Beds (Turonian to Coniacian)
LSK Lower part of Seewen Formation (Upper Albian to Upper Cenomanian)
UFB Upper part of Freschen Beds (Lower Albian to Upper Albian)

Fig. 33. *Archaeocretacea* Zone Götzis Bed.

A. Götzis Bed (g) in an inverted sequence near Götzis. Note the sharp contact to underlying Seewen sediments (s). Boundary to the overlying Seewen micrites (m) is transitional. Bed thickness approximates 1.8 m (cf. Heim 1958).

B. *Thalassinoides* burrows on the bedsole of the Götzis Bed. The burrows are infilled with glauconitic sandstone and penetrate underlying Seewen sediments (s) (surface parallel to bedsole)

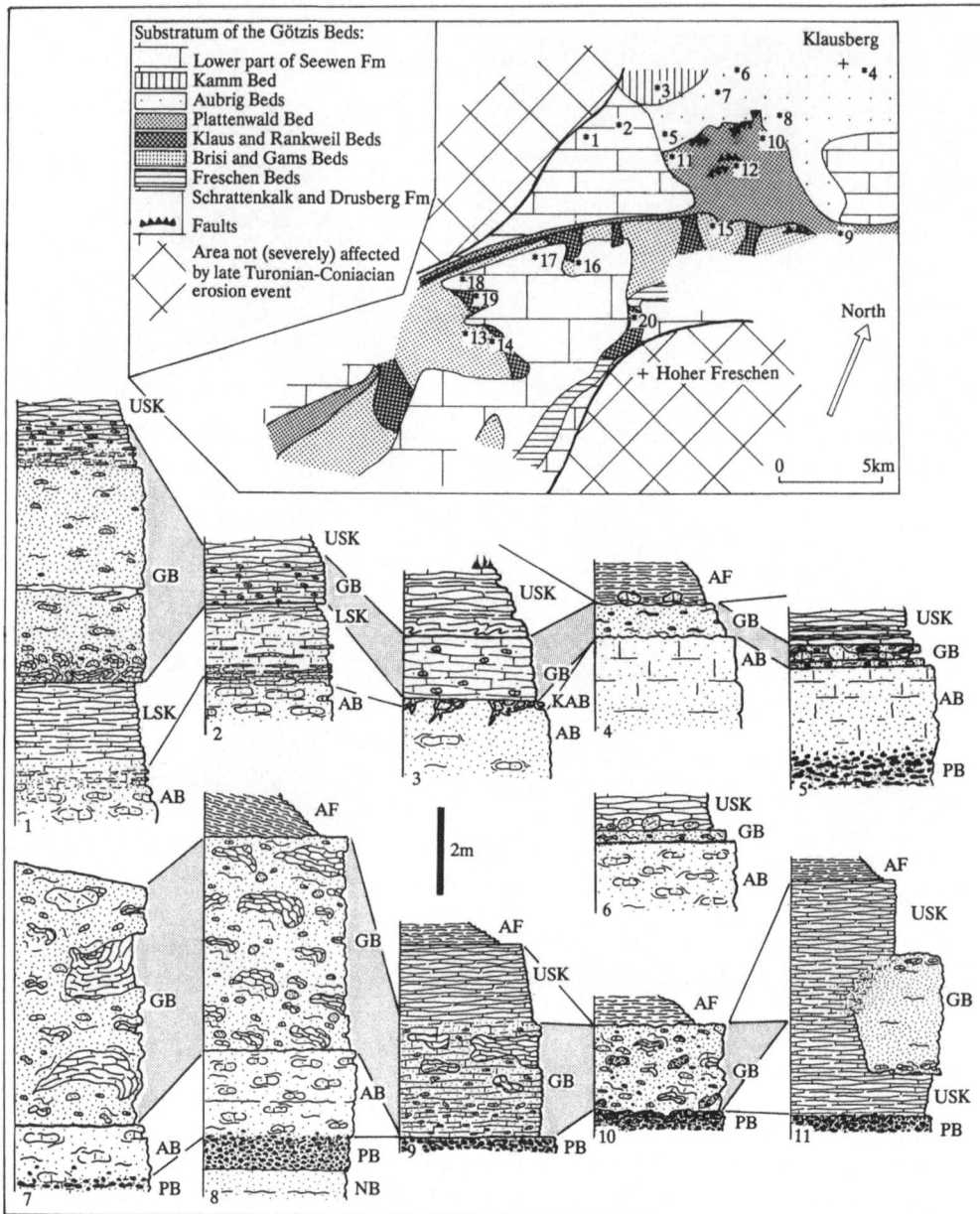

Fig. 34. Late Turonian to Coniacian.
Palinspastic map shows the distribution area of the Upper Turonian to Conia-
cian Götzis Beds. This area is limited to the eastern part of the inner shelf, the
Rankweil Ramp, and adjacent areas. Note that this area is complementary to
the *archaeocretacea* zone Götzis Bed distribution area, displayed in Fig. 32.
The Götzis Beds substratum consists of sediments of the Seewen, Garschella,

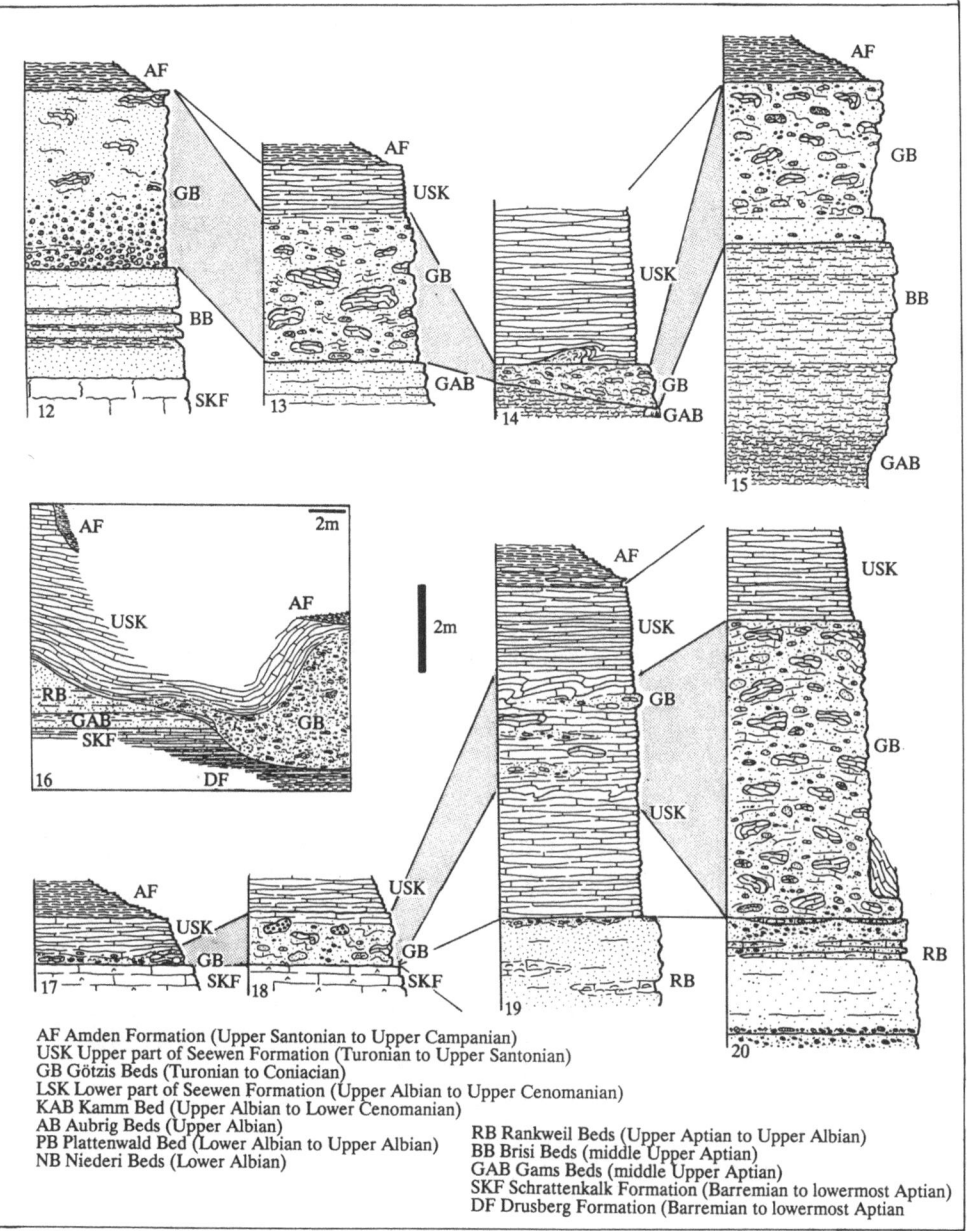

AF Amden Formation (Upper Santonian to Upper Campanian)
USK Upper part of Seewen Formation (Turonian to Upper Santonian)
GB Götzis Beds (Turonian to Coniacian)
LSK Lower part of Seewen Formation (Upper Albian to Upper Cenomanian)
KAB Kamm Bed (Upper Albian to Lower Cenomanian)
AB Aubrig Beds (Upper Albian)
PB Plattenwald Bed (Lower Albian to Upper Albian)
NB Niederi Beds (Lower Albian)

RB Rankweil Beds (Upper Aptian to Upper Albian)
BB Brisi Beds (middle Upper Aptian)
GAB Gams Beds (middle Upper Aptian)
SKF Schrattenkalk Formation (Barremian to lowermost Aptian)
DF Drusberg Formation (Barremian to lowermost Aptian)

Schrattenkalk, and Drusberg Formations, reflecting the varying amounts of erosion that occurred during the late Turonian-Coniacian episode. During this period, the area of the Rankweil Ramp was subjected to large-scale failures. Note the heterogeneity of the Götzis Beds in the sections. Correlation of the Upper Turonian-Coniacian Götzis Beds is indicated by dots

The restricted distribution area and the uniform appearance of the *archaeocretacea* Zone Götzis Bed suggest that this bed resulted from a single gravity flow (debris or high-density turbidity flow), which originated from a point source or a line source of limited extent in a proximal part of the inner shelf. It is not clear whether the glauconitic sands represent reworked Garschella sediments or more contemporaneous sediments, otherwise unknown from the helvetic shelf. The angular quartz grains (max. 0.6 mm; mean grain size 0.25 mm), and the glauconite particles (10-20%; grain size similar to quartz particles) are quite reminiscent of Brisi Beds, favoring therefore the possibility of reworking from the Garschella Formation.

The Götzis Bed gravity flow eroded the surface of the underlying Seewen sediments (with a tendency to erode slightly deeper toward the Rankweil Ramp; Fig. 32). The eroded, unconsolidated Seewen micrites have partly been included into the sandy gravity flow, and partly resuspended and subsequently redeposited on top of the sandy bed (indicated by an allochthonous globotruncanid fauna of *cushmani* and *archaeocretacea* Zone ages, present at the base of overlying Seewen sediments).

Along the Rankweil Ramp, sediment accumulation started again during the *archaeocretacea* Zone. However, pelagic micrites of this age are only locally preserved, especially within scoured holes of subjacent Plattenwald Bed occurrences, where they have been protected against subsequent erosion events (Fig. 28).

2.10 Middle Turonian

Autochthonous middle Turonian sediments are preserved in approximately the same areas, where the *archaeocretacea* Zone Götzis Bed is preserved. In areas affected by the Turonian-Coniacian boundary erosive event, sediments of this and older ages were subsequently incorporated into Upper Turonian-Coniacian Götzis Beds (Sect. 2.11; Figs. 32 and 34).

Both autochthonous and redeposited sediments from this period (*helvetica* Zone) consist of typical Seewen micrites, indicating the continuing dominance of the pelagic sedimentary regime across the entire helvetic shelf.

2.11 Late Turonian to Coniacian

In the late Turonian-Coniacian period (*sigali*, *primitiva*, and early *concavata* Zones; <2 my; Fig. 2), large areas of the Vorarlberg inner shelf and the Rankweil Ramp experienced a strong erosive and redeposition event, which produced the following modifications within previously deposited sediments (Fig. 34):

1. In the central and eastern part of the Vorarlberg inner shelf, this erosive event erased most of the Seewen Formation (sections 1-3 in Fig. 34; Fig. 38) and leveled down to the Aubrig Beds (in proximal areas; sections 4-8 in

Fig. 34) and the Plattenwald Bed (in distal areas; sections 9-11 in Fig. 34). Numerous subvertical faults were generated or reactivated during this event (Fig. 34). Their orientation is approximately parallel to the Rankweil Ramp and their maximum vertical offset amounts to 30 m. Faulting was accompanied by breccia formation, which accumulated in fault talus fans up to 5 m thick (section 12 in Fig. 34; Fig. 35). Faults with both northfacing and southfacing escarpments occur, thus enclosing fault-bounded uplifts, which are parallel to the Rankweil Ramp. This structural configuration was probably induced by normal faulting and the relative elevation of more rigid blocks within relatively incompetent sediment masses, related to coeval failure of the nearby Rankweil Ramp (see below; Figs. 34 and 49; Coleman and Prior 1980; Coleman 1988; Sect. 3.7.3).

2. The late Turonian-Coniacian erosive episode had a large impact on sediments along the Rankweil Ramp. Major parts of the Garschella Formation disappeared; in a large area exposed in western Vorarlberg, Garschella sediments became completely eroded and Schrattenkalk, and Drusberg sediments reexposed (Fig. 34, sections 13-20).

3. In areas close to the affected area, a hiatus within the Seewen Formation including the *sigali* and sometimes the *primitiva* Zone relates to this event. In more distal areas, the deposition of Seewen micrites continued.

This erosive episode produced a large amount of reworked sediments, preserved in the above mentioned areas (Fig. 34; Upper Turonian to Coniacian **Götzis Beds**; max. 30 m; Föllmi 1981, 1986; Föllmi and Ouwehand 1987). The Upper Turonian-Coniacian Götzis Beds consist of debris flow deposits, which

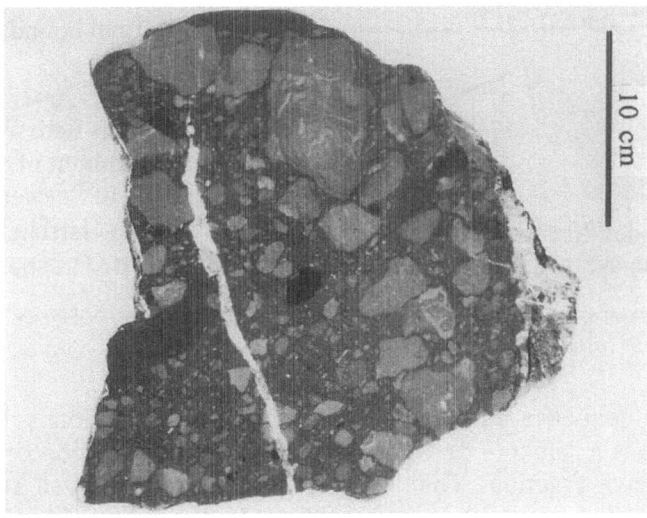

10 cm

Fig. 35. Upper Turonian-Coniacian Götzis Beds.
Polished slab of breccia containing Schrattenkalk and subordinate Seewen components, embedded in a glauconitic sandstone (derived from the Garschella Formation). This breccia is part of a fault talus

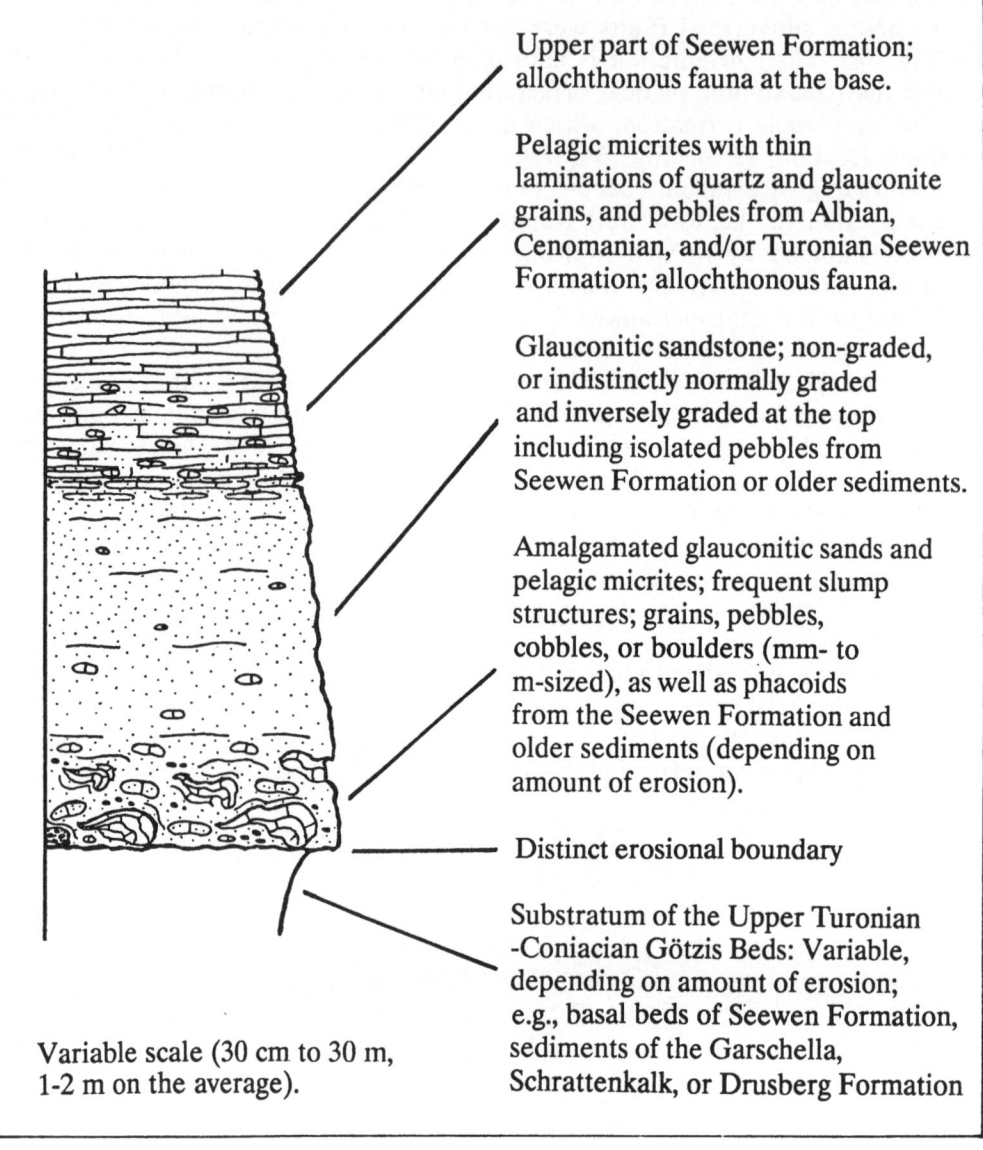

Upper part of Seewen Formation; allochthonous fauna at the base.

Pelagic micrites with thin laminations of quartz and glauconite grains, and pebbles from Albian, Cenomanian, and/or Turonian Seewen Formation; allochthonous fauna.

Glauconitic sandstone; non-graded, or indistinctly normally graded and inversely graded at the top including isolated pebbles from Seewen Formation or older sediments.

Amalgamated glauconitic sands and pelagic micrites; frequent slump structures; grains, pebbles, cobbles, or boulders (mm- to m-sized), as well as phacoids from the Seewen Formation and older sediments (depending on amount of erosion).

Distinct erosional boundary

Substratum of the Upper Turonian -Coniacian Götzis Beds: Variable, depending on amount of erosion; e.g., basal beds of Seewen Formation, sediments of the Garschella, Schrattenkalk, or Drusberg Formation

Variable scale (30 cm to 30 m, 1-2 m on the average).

Fig. 36. Ideal sequence in Upper Turonian-Coniacian Götzis Beds. The sequence embodies a "fluxo"- or "megaturbidite", i.e., deposition of a debris flow with a suspensive fraction. The "base" of superjacent Seewen sediments consists of redeposited material, representative of the finest fraction within the megaturbidite. The completeness of the megaturbidite sequence in the Götzis Beds depended mainly on topographic features, the availability and physical properties of the included redeposited sediments, as well as the proximity to the source area

are commonly topped by turbidites ("fluxo-" or "megaturbidites"; Kuenen 1958; Dzulynski et al. 1959; Bouma 1987; Bourrouilh 1987; Figs. 36 and 37). On the inner shelf, mainly sheet-like debris flows developed and only a minority occur in channels (small; max. 5-10 m wide and 1-3 m deep; e.g., section 11 in Fig. 34; compare also Fig. 38). Along the Rankweil Ramp, the majority of the Götzis Beds debris flows are channelized (max. 50 m wide and 10 m deep; e.g. section 16 in Fig. 34).

Götzis debris flow deposits consist mainly of nongraded or indistinctly normally to inversely graded glauconitic sandstones. At the base, these sands are usually slump-folded, interstratified or amalgamated with Seewen micrites, and contain mm to m-sized lithoclasts as well as phacoids from the Seewen, Garschella, Schrattenkalk, and Drusberg Formation (Fig. 37). The composition of these lithologies depended on the level of erosion and the cohesiveness of the involved sediments. The ubiquitous presence of Garschella lithoclasts (e.g., reworked phosphatic fossils) suggests that the matrix of glauconitic sands is derived from eroded Garschella sediments as well. At the top, the sandstones are commonly covered with Seewen micrites, which include an allochthonous globotruncanid fauna at the base (Fig. 36).

In peripheral areas, where erosion was restricted to the Seewen sediments and underlying formations have not been involved, mud and turbiditic flows of pure pelagic micrites developed. The micritic calciturbidites commonly include a mixed fauna, which spans several age zones, whereas the mud flow deposits contain completely reworked faunas of one age zone. These types of gravity flow deposits are commonly difficult to identify, especially when they occur within a sequence of "autochthonous" Seewen micrites.

In general, the character and composition of the upper Turonian-Coniacian Götzis Beds depended on (1) the level of erosion in the source area; (2) the degree of cohesion in the involved sediments (noncohesive glauconitic sands versus cohesive micrites; the availability of lithified sediments); (3) the topography of the area; and (4) proximity to the source area.

The ages of the youngest redeposited sediments included in the Götzis Beds (late Turonian and Coniacian) and the overlying, "autochthonous" Seewen sediments (Coniacian, Santonian) suggest a late Turonian-Coniacian age for this

Fig. 37 (next page). Polished slabs of the Upper Turonian-Coniacian Götzis Beds.
A. Phacoid of Middle Turonian Seewen micrites, slump-folded around a core of glauconitic sandstone (derived from the Garschella Formation). Note the laminations in the micrites, probably caused by the release of internal shear stress, and the injections of glauconitic sand into the micrites. The phacoid has been found in a 30-m-thick debris-flow deposit, consisting essentially of glauconitic sandstone (Föllmi 1981; Plate 1, Fig. 1).
B. Phacoid of Middle Turonian Seewen micrites, embedded in an amalgamate of Seewen micrites and Garschella glauconitic sands. Debris-flow deposition at the base of section 1 in Fig. 34 (Föllmi 1981; Plate 2, Fig. 2; cf. Voigt 1962)

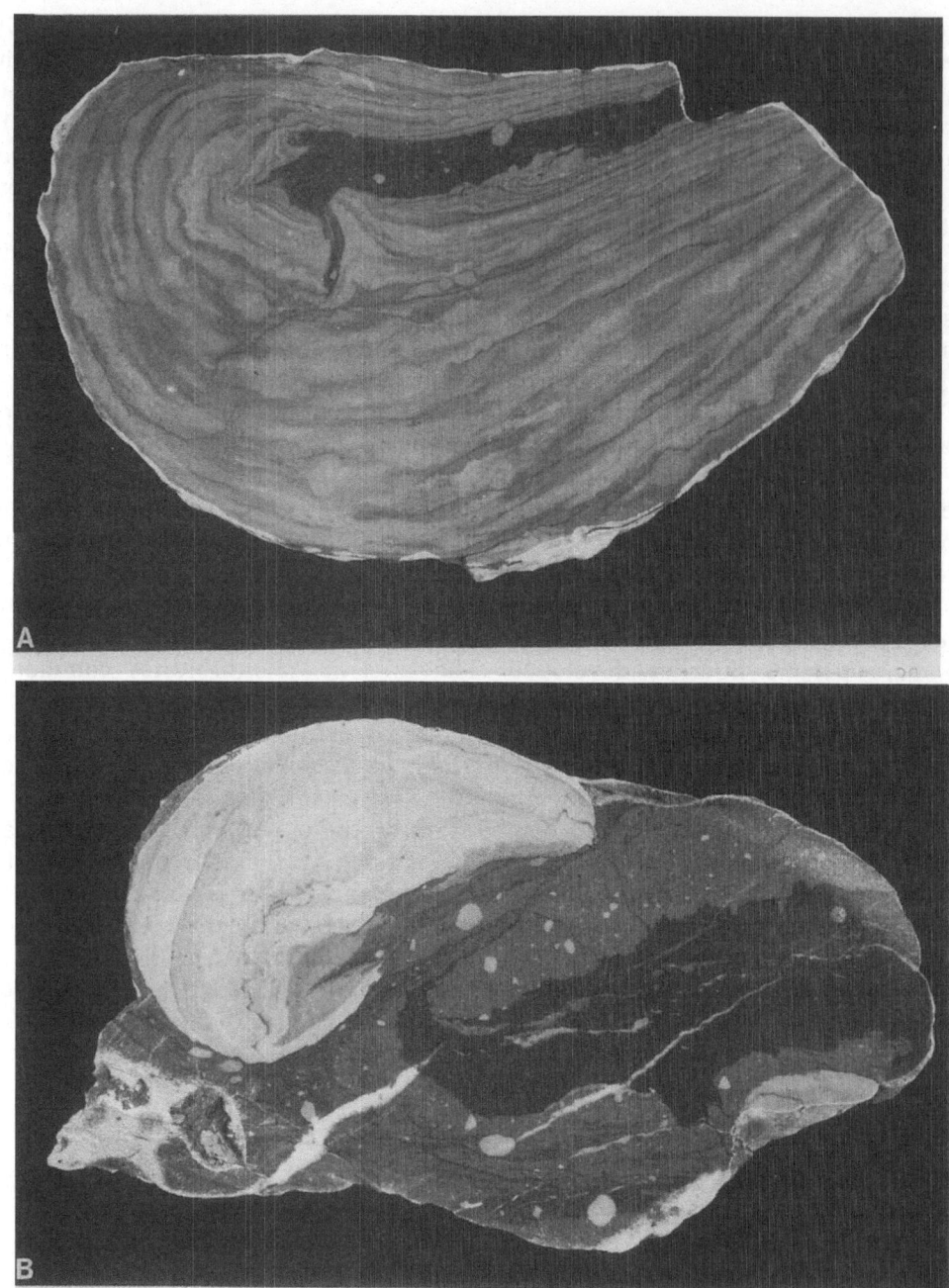

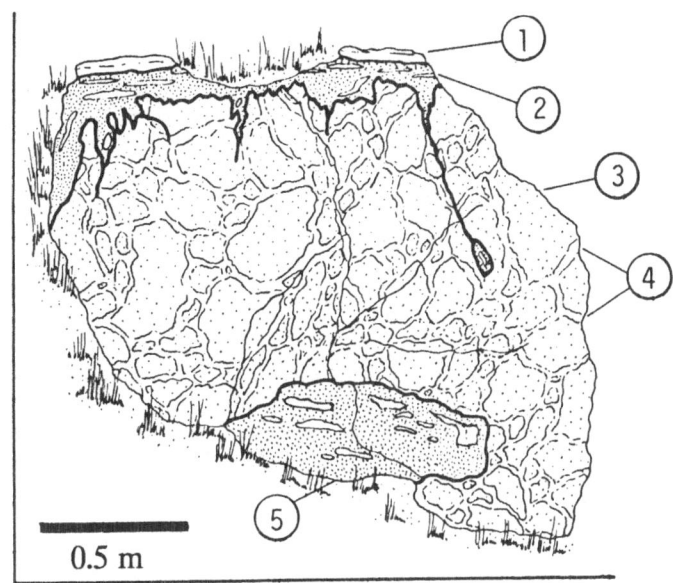

Fig. 38. Upper Turonian-Coniacian Götzis Beds.
Götzis Beds consisting of glauconitic sands including micritic lenses form a
thin veneer (2) on top of Kamm Bed sediments (3, 4), separated by a distinct,
erosional boundary. The redeposited sediments penetrate the underlying Kamm
Bed along subvertical and subhorizontal fractures (5). The Götzis Beds are covered by sediments of the Seewen Formation (1). This outcrop is located near
Hof, in the Bregenzer Ache valley

episode. In many cases dating is difficult, because short-ranged globotruncanids of this period are rare and difficult to determine in thin sections, whereas
larger range globotruncanids such as *Marginotruncana coronata* (BOLLI) pre-
dominated (Fig. 41; Föllmi 1986).

The late Turonian-Coniacian erosive event documented in the massive erosion and redeposition of sediments and confined to the distal inner shelf and
the Rankweil Ramp, is interpreted as a failure event, in which the Rankweil
Ramp and the adjacent unstable inner shelf area collapsed. The extensive failure was probably due to oversteepening of the ramp (differential subsidence
of the inner and outer shelf since early Aptian times; cf. sequences of Figs. 6,
10, 16, 20, and 30; Sect. 3.2.2).

Coeval Götzis Beds are known from Allgäu (southeastern F.R.G.), Swit-
zerland, and southeastern France (e.g., Fichter 1934; Heim and Seitz 1934;
Bolli 1944; Bentz 1948; Keller 1983; Föllmi 1981; Föllmi and Ouwehand 1987).
Their discrete appearance in distal inner shelf localities indicates that the late
Turonian-Coniacian boundary event had a larger, though nonuniform, impact
on the helvetic shelf (several coexisting "epicenters").

Fig. 39. Southward dipping Coniacian to Upper Santonian sediments of the Seewen Formation, covering Brisi Beds with a 25 degree unconformity (close to section 9 in Fig. 7)

2.12 Late Coniacian to Late Santonian

During late Coniacian to late Santonian times, deposition of the pelagic Seewen Formation continued. Frequent sediment perturbations occur within this upper part of the Seewen Formation, ranging from winnowed layers and low-angle truncations (Figs. 39, 40, and 41) to m-thick mud flow deposits. The hiatuses and event beds indicate prevailing dynamic conditions on the helvetic shelf during this time interval.

The base of the Coniacian-Upper Santonian Seewen Formation is commonly represented by an angular unconformity, probably related to the late Turonian-Coniacian event (Fig. 39; e.g., section 12 in Fig. 19).

Larger areas within the eastern part of the Vorarlberg inner shelf and along the Rankweil Ramp lack Seewen sediments of this time interval. This is probably due to zero net sediment accumulation rates and especially to erosion prior to (?) and during deposition of the basal Amden Formation beds (see following Sect.).

Fig. 40. Internal low angle truncations within Coniacian to Upper Santonian sediments of the Seewen Formation (close to section 4 in Fig. 7)

2.13 Late Santonian to Early Campanian

The late Santonian interval is characterized by another major change in sediment deposition within the helvetic shelf area. On the inner shelf and on proximal parts of the outer shelf, the pelagic Seewen regime ceased and a detritus-influenced depositional system developed, consisting of silty and sandy marls and muds (Upper Santonian to Upper Campanian **Amden Formation**; max. 400 m; Oberhänsli 1978). In distal parts of the outer shelf, deposition of Seewen micrites continued (into the Campanian; Weidich 1987).

The base of the Amden Formation is marked by a distinct, commonly an-

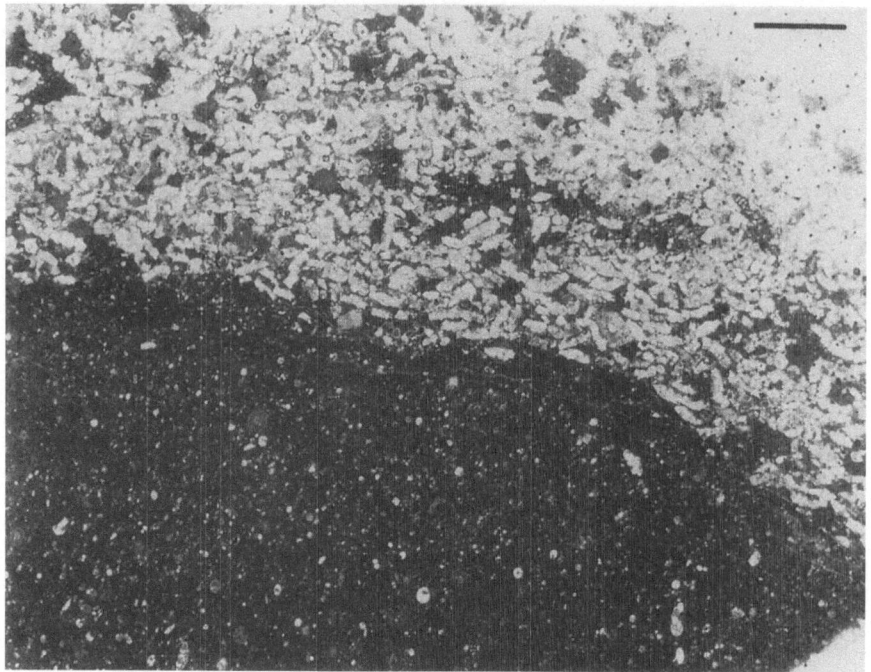

Fig. 41. Thin-section photomicrograph of Coniacian to Upper Santonian sediments of the Seewen Formation (<u>bar</u> = 1 mm). Local concentrations of globotruncanid foraminifera [predominantly *Marginotruncana coronata* (BOLLI)] is probably due to winnowing (Seewen Formation in ultrahelvetic nappe remnant of the Hohe Kugel area; section 17 in Fig. 19)

Fig. 42. Amden Formation.
A. Mega-"phacoid" in a basal sequence of the Amden Formation. This phacoid consists of Upper Turonian to Coniacian Götzis Beds and Coniacian to Santonian Seewen sediments, which are slump-folded around a core of Amden Formation (<u>arrow</u>). Note the adjacent steep Amden beds and the high-angle unconformity between these beds and superjacent horizontal Amden beds. It is unclear whether the phacoid of reworked, older sediments embodies a solitary sedimentary slide, or the whole complex of older redeposited sediments <u>and</u> the steep Amden beds belong to a larger olistostrome-type gravity flow (cf. Föllmi 1981, Fig. 2).
B. Diagrammatic cross-section

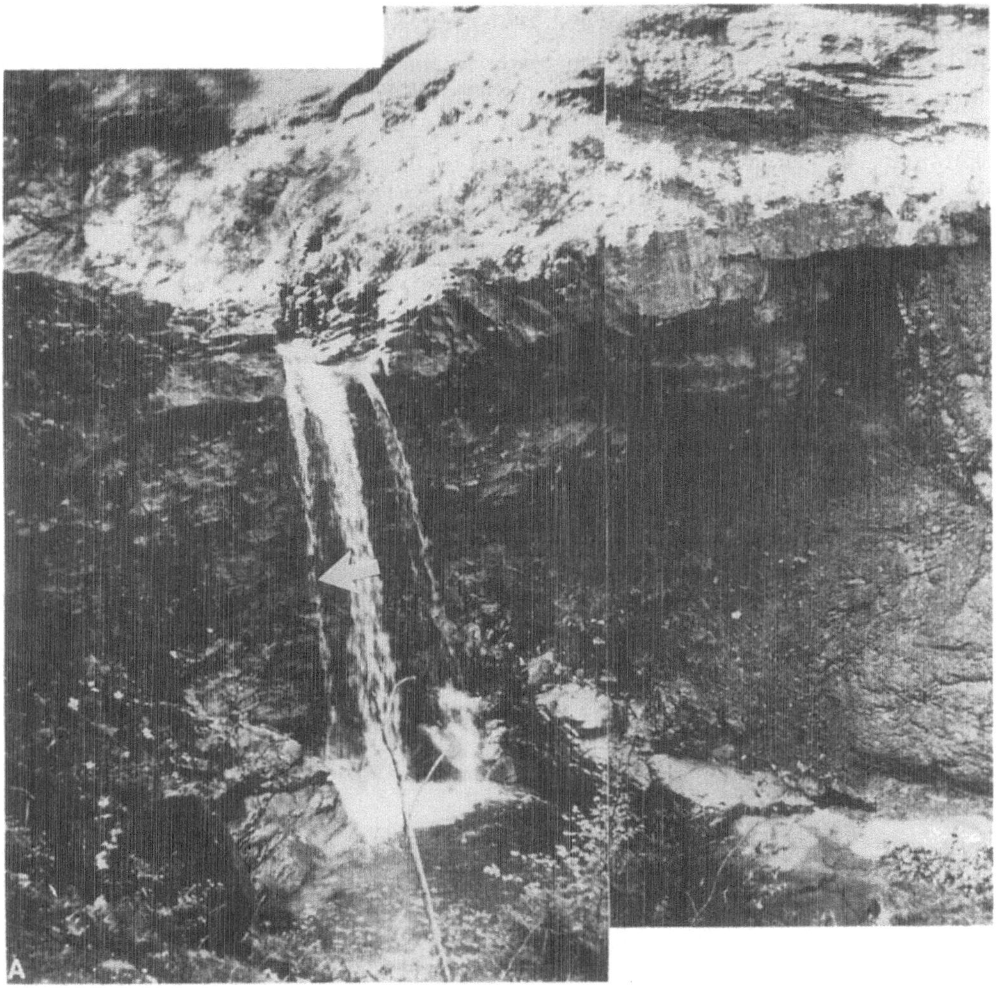

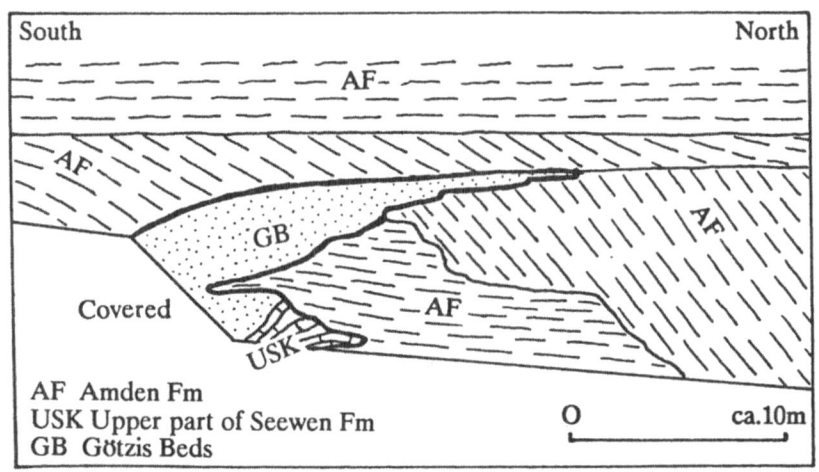

South
North
AF
AF
GB
AF
Covered
AF
USK
AF Amden Fm
USK Upper part of Seewen Fm
GB Götzis Beds
O
ca.10m
B

gular unconformity (e.g., section 8 in Fig. 34). Prior to or during the deposition of basal Amden sediments, an erosive event erased major parts of the underlying Seewen Formation, especially in the eastern areas of the Vorarlberg inner shelf and along the Rankweil Ramp. In localities along the Rankweil Ramp, basal Amden sediments infill channels incised into Coniacian-Upper Santonian Seewen sediments (e.g., section 16 in Fig. 34). Sedimentary slides of the Seewen Formation, Götzis Beds, and older sediments are very common in basal Amden sediments throughout the Vorarlberg Amden distribution area (e.g., section 8 in Fig. 7; Fig. 42; Föllmi 1981, 1986). However, it is difficult to infer whether these deposits represent genuine sedimentary slides or megaphacoids within an olistostrome-type gravity flow (Fig. 42). The sharp, commonly unconformable base of the Amden Formation, the channelized basal beds perpendicular to the Rankweil Ramp, and the concomitant erosive phase, involving underlying Seewen sediments, favor the latter explanation, suggesting that larger basal Amden sequences within the entire Vorarlberg distribution area in fact are large-scale gravity flow deposits.

3 Mechanisms Controlling the Formation of the Barremian-Campanian Vorarlberg Helvetic Shelf Sequence

3.1 Introduction

What was the character of the driving forces, which contributed to the generation of the sequence of shallow-water carbonates, phosphatic and glauconitic sediments, pelagic limestones, and detrital sediments on the helvetic shelf, recording a cycle of deepening and shallowing upward, and how did these forces interact? In the following sections, topographic, paleontologic, paleoceanographic, and tectonic phenomena, and their possible bearing upon the configuration of the Barremian-Campanian sediments will be examined, in an attempt to unravel the complex history of this Mid- and lower Upper Cretaceous sequence and to trace a development in sedimentary patterns typical of the Mid-Cretaceous interval.

3.2 Sea Bottom Topography and Differential Subsidence

The topography of the helvetic shelf, part of a passive margin, represented an important factor in controlling sediment distribution. Reciprocally, the topography itself was determined by the lithologic character, sediment accumulation rates, as well as compaction rates of the shelf sediments, in addition to the tectonic variables.

It can be shown qualitatively that differential subsidence, i.e., different rates of compaction and subsidence of the inner and outer shelf, played a major role in the Mid- to early late Cretaceous evolution of the helvetic shelf morphology, especially in the development of the Rankweil Ramp.

Funk (1985) provided quantitative information on general subsidence rates of the eastern Swiss helvetic inner shelf, based on geohistory diagrams. His dates can be transferred to the Vorarlberg inner shelf as a first-order estimation (approximately 10-13 m/my for the distal inner shelf during the Aptian-Santonian interval).

3.2.1 Narrowing and Steepening of the Carbonate Platform in the Eastward Direction

Facies belts within the Schrattenkalk and Garschella Formations narrow toward the east (Figs. 4, 7, 15, and 19; Zacher 1973; Bollinger 1986). Net sediment accumulation rates recorded in these formations decrease toward the eastern inner shelf. These two phenomena are interpreted as the result of increased hydrodynamic activity toward the east, as is exemplified in eastern Vorarlberg by the absence of the Luitere Bed (Fig. 4), the eastward wedging out of the Sellamatt Bed sandbody (Fig. 19), the integration of the Klaus Beds into younger condensation processes of the Plattenwald Bed (Figs. 19 and 28), the occurrence of hydraulic erosional "tunneling" within the Plattenwald Bed (Fig.

28), and the presence of major erosion during the Upper Turonian-Coniacian event (Fig. 34). These features are attributed to an eastward narrowing and steepening of the underlying Schrattenkalk platform, which is also indicated by a northward withdrawal of the platform toward the east (Figs. 4 and 57; bight-shaped platform margin).

3.2.2 Transformation of the Carbonate Platform Margin from a Gentle Homocline into a Distally Steepened Ramp (Rankweil Ramp)

Near the Barremian-Aptian boundary, the Schrattenkalk carbonate platform reached the zenith in its development. At this time, the platform was connected with the muddy outer shelf through a gentle homoclinal ramp, which apparently lacked significant channels (Sect. 2.2; Read 1985; Bollinger 1986).

First indications of steepening of the platform margin emerged during the early and early late Aptian, as the platform drowned (Sect. 3.8). Small but distinct channels appeared along the ramp, as well as in proximal areas of the outer shelf. These channels contain eroded inner shelf "Upper Orbitolina Beds" and Schrattenkalk sediments, which were transported downslope, to be incorporated into the outer shelf Mittagspitz Formation (Sect. 2.3.1).

During this period, differential subsidence on the inner shelf caused the incorporation of the distal platform part in western Vorarlberg into the transition zone to the outer shelf (Figs. 4, 7, and 15), as is indicated by the onlap of the Luitere, Gams, and Rankweil Beds onto the top of this marginal platform area. In eastern Vorarlberg, these sediments are limited to proximal outer shelf settings beyond the platform (Sects. 2.3.1, 2.4.1, and 2.5.1). During early Aptian time, the western Vorarlberg distal platform part was subjected to reinforced subsidence. Sudden facies changes and the presence of thin debris flows within the Luitere Bed, onlapping onto the platform area, are an additional indication of the breakup of this distal platform part (Fig. 4). The anomalous strong subsidence was probably due to the compaction of particularly thick, marly Drusberg sequences below this marginal platform area.

In middle late Aptian time, the platform margin still functioned as an area of broad facies transitions between the inner and outer shelf, as is indicated by its cover of Gams Beds (transitional between inner shelf siliciclastic Brisi Beds and outer shelf muddy and marly Freschen Beds). Distinct, laterally shifting channel and fan systems developed along the ramp, carrying inner shelf sediments toward proximal areas of the outer shelf (Figs. 7 and 11; Sect. 2.4.1).

Obviously, the homoclinal platform margin underwent a slow and steady process of steepening during Aptian time. This process was most probably induced or at least catalyzed by different compaction rates of the inner shelf with a foundation of platform carbonates and sandstones, and the outer shelf with a foundation of thick mud and marl sequences (e.g., Joseph et al. 1986).

In latest Aptian time, the process of steepening accelerated dramatically. During this period, large and stable channels were incised into the ramp. Their magnitude (max. 25 m deep; several 100 m wide) greatly exceeds that of older channels. The ramp was transformed from a facies transition zone into a distinct facies boundary along its upper limit (e.g., Klaus Beds and Rankweil Beds; Plattenwald Bed and Rankweil Beds; Figs. 2, 15, 19, 30, and 49; Sects. 2.5.1 and 2.6.1).

By the end of the Aptian, the homoclinal transition zone between the inner and outer shelf had changed into a distally steepened ramp (designated as Rankweil Ramp). In eastern Vorarlberg, the Rankweil Ramp is identical to the former carbonate platform margin; in western Vorarlberg, it is located within the former carbonate platform, due to the above mentioned incorporation of a distal platform area into the transitional zone toward the outer shelf during early Aptian time.

The sudden increase in steepening may be attributed to reinforced differential subsidence between the inner and outer shelf, which has probably been evoked by the onset of a transpressional regime in the helvetic area (Sect. 3.7.3).

During Albian, Cenomanian, and Turonian times, the Rankweil Ramp represented a stable topographic feature on the helvetic shelf not subjected to major changes.

Near the Turonian-Coniacian boundary, the Rankweil Ramp and adjacent areas experienced massive failure, which was probably related to oversteepening of the rampzone. The process of failure led to a certain degree of adjustment between the inner and outer shelf (Sect. 2.11). From the Coniacian onward, the dilated Rankweil Ramp lacked its pronounced slope (Fig. 49).

3.3 Ammonoid Paleobiogeography

Ammonoids are common in phosphatic beds of the Garschella Formation (Luitere, Twäriberg, Durschlägi, Wannenalp, Plattenwald, and Kamm Beds) and appear as well in redeposited sediments, which include eroded material from the phosphatic beds (Klaus, Rankweil, and Götzis Beds). These fossils provide a biostratigraphic control within the Garschella Formation with an optimal resolution of approximately 0.5 my (Fig. 2; Föllmi 1986, 1989). Beyond that, they render specific biogeographic information with respect to their distribution areas (Fig. 43):

1. Early to early late Aptian ammonoids predominantly include species which are limited to, or abundant in the tethyan area [*Phylloceras (Hypoph.)*, *Beudanticeras (Zürcherella)*, *Puzosia (Melchiorites)*, *Colombiceras*, *Cheloniceras*, *Parahoplites*; Fig. 43: 1-2; Föllmi 1986, 1989].
2. Latest Aptian and earliest Albian ammonoids indicate an important change in paleogeographic conditions. Characteristic tethyan species are reduced to a minority and Boreal or pan-European species prevail (e.g., *Hypacanthoplites*, *Leymeriella*; Fig. 43: 3-4).
3. Late early and middle Albian ammonoids include a suite of typical tethyan species (e.g., *Phylloceras*, *Lytoceras*, *Protetragonites*, *Pictetia*, *Kossmatella*, *Jauberticeras*, *Tetragonites*, *Hamites*, *Anisoceras*; Fig. 43: 5-11), which are

associated with significant Boreal species (*Otohoplites*, *Hoplites*, *Anahoplites*; Fig. 43: 12).

4. Late Albian ammonoids are predominantly cosmopolitan: (e.g., *Hysteroceras*, *Prohysteroceras*, *Mortoniceras*, *Dipoloceras*, *Neophlycticeras*, *Stoliczkaia*, *Labeceras*; Fig. 43: 13-15).

The Vorarlberg ammonoid assemblage shows a specific tethyan-cosmopolitan character. Boreal influences are limited to the Aptian-Albian boundary (strong) and early to middle Albian (weaker). Intriguingly, the Vorarlberg tethyan ammonoids exhibit strong affinities to coeval eastern European faunas, especially in the early late Aptian. Most Vorarlbergian species of *Colombiceras*, *Cheloniceras*, and *Parahoplites* are established in, or described from the southern Soviet Republics. In contrast, minor affinities exist to the nearby classical Aptian and Albian faunas from southeastern France (e.g., areas around Apt and Escragnolles: different species of *Cheloniceras* and *Colombi-*

Fig. 43. Characteristic ammonoids of the Garschella Formation (Föllmi 1986, 1989).

1. *Colombiceras tobleri* (JACOB & TOBLER); Luitere Bed; early late Aptian
2. *Cheloniceras subnodosocostatum* (SINZOW); Luitere Bed; early late Aptian
3. *Hypacanthoplites subrectangulatus* (SINZOW); Rankweil Beds; latest Aptian
4. *Leymeriella* (*Neoleym.*) *seitzi* sp. nov.; Plattenwald Bed; early Albian
5. *Phylloceras* (*Hypophyll.*) *subalpinum ellipticum* KOSSMAT; Plattenwald Bed; middle Albian
6. *Pictetia oberhauseri* sp. nov.; Plattenwald Bed; middle Albian
7. *Protetragonites aeolus aeoliformis* (FALLOT); Plattenwald Bed; middle Albian
8. *Jauberticeras* aff. *latericarinatum* (ANTHULA); Plattenwald Bed; middle Albian
9. *Kossmatella ventrocincta ventrocincta* (QUENSTEDT); Plattenwald Bed; middle Albian
10. *Hamites maximus* SOWERBY, Plattenwald Bed, middle Albian
11. *Anisoceras* (*Anis.*) *arrogans* (GIEBEL); Plattenwald Bed; middle Albian
12. *Hoplites* (*Hopl.*) *escragnollensis* SPATH; Plattenwald Bed; middle Albian
13. *Labeceras* (?) *collignoni* sp. nov.; Wannenalp Bed; late Albian
14. *Hysteroceras crassicostatum* (JAYET); Wannenalp Bed; late Albian
15. *Mortoniceras* (*Deiradoceras*) aff. *exile* (VAN HOEPEN); Plattenwald Bed; late Albian

1-2. Tethyan species, related to eastern European assemblages
3-4, and 12. Boreal or pan-European species
5-11. Tethyan species
13-15. Cosmopolitan species
1-7, 9-12, 14-15: 0.8 x natural size; 8, 13: 1.6 x natural size

ceras; abundant *Aconeceras, Gargasiceras, Lyelliceras, Mojsicovicsia*, and *Brancoceras*, which are absent or very rare in Vorarlberg). These similarities and differences to coeval faunas are probably depth-related (e.g., prevalence of lyto- and phylloceratid ammonoids in deeper waters), but above all current-related. A westbound geostrophic current system along the northern Tethys margin was responsible for the strong affinities between Vorarlbergian and eastern European assemblages (Fig. 44).

The observed invasion of Boreal ammonoids in the latest Aptian was most probably made possible by the documented late Aptian opening of the Anglo-Parisian Basin seaway to the Boreal sea, *via* the London-Brabant-Ardennes Massif (Fig. 44; Juignet et al. 1973; Föllmi 1986; Bréhéret et al. 1986). The outspoken cosmopolitan character of late Albian ammonoids was related to continuing sea level rises, which facilitated ammonoid spreading and communication.

3.4 Currents

One of the key observations of Arn. Heim is his notion of the current-dominated character of the helvetic shelf during Mid-Cretaceous times and its bearing on the helvetic sediment distribution and configuration, in general, and on the origin of discontinuities and formation of phosphatic beds, in particular (Heim 1924, 1934, 1946, 1958; Heim and Seitz 1934). Heim's conception of a current system along the northern Tethys margin has been reconfirmed by several investigators (Luyendyk et al. 1972; Fischer 1981, Delamette 1985, 1988a,b; Föllmi 1986; Bréhéret et al. 1986; Seidov 1986; Ouwehand 1987). It provides a powerful tool for deciphering the Aptian-Cenomanian sedimentary column.

What is the Vorarlberg evidence for this current system, and how precisely can it be reconstructed?

The Garschella ammonoids give a first indication of current activity along the northern Tethys margin, as shown above. They point to a northern tethyan, westbound current system throughout Aptian-early Cenomanian times, as well as to a Boreal-Tethys current from the latest Aptian onwards (Fig. 44).

The sediments of the Garschella Formation show a conspicuous zonation, approximately parallel to the carbonate platform margin, respectively to the Rankweil Ramp (Figs. 4, 15, 19, 49, and 51; Föllmi 1986; Ouwehand and Föllmi 1987):

Zone A. A proximal zone with moderate sediment accumulation rates, in which glauconitic sandstones and marls have been deposited (approximately 10-30 m/my; "Upper Orbitolina Beds", Niederi, Sellamatt, and Aubrig Beds).

Zone B. A zone with very low, zero, or negative sediment accumulation rates, in which erosion, nondeposition, and/or the formation of thin phosphatic beds occurred (shifting between the intermediate and distal inner shelf; max. 20 cm/my; Luitere, Twäriberg, Durschlägi, Wannenalp, Plattenwald,

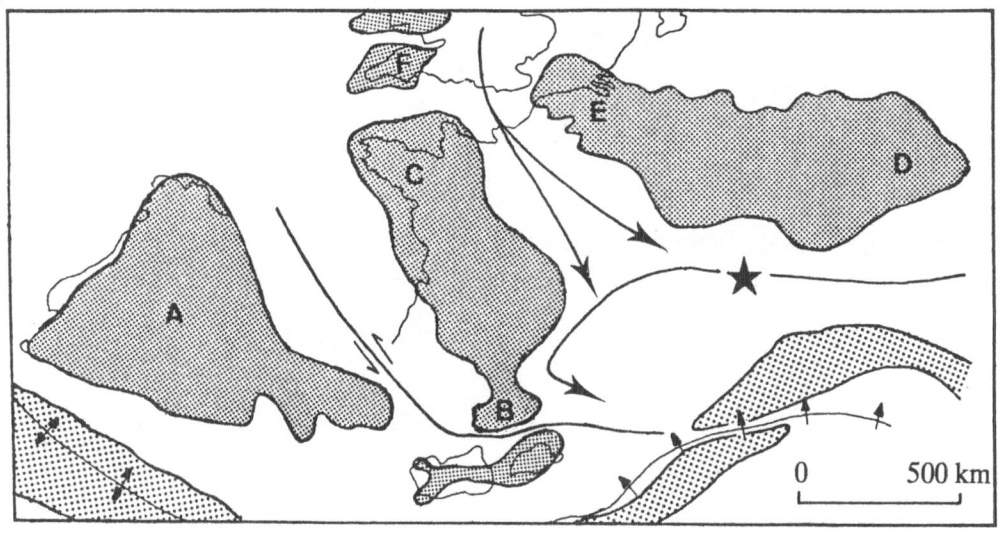

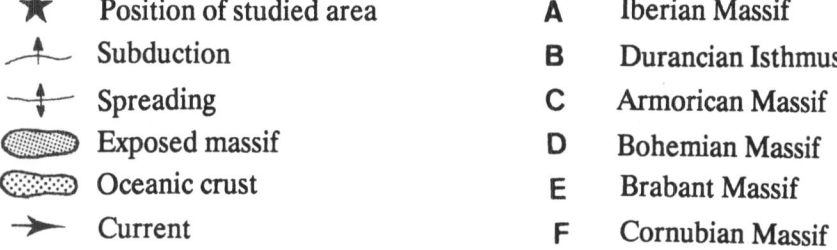

★	Position of studied area	**A**	Iberian Massif
	Subduction	**B**	Durancian Isthmus
	Spreading	**C**	Armorican Massif
	Exposed massif	**D**	Bohemian Massif
	Oceanic crust	**E**	Brabant Massif
	Current	**F**	Cornubian Massif

Fig. 44. Early Albian paleogeographic map of western Europe (after Ziegler 1982, 1988; Dercourt et al. 1985, 1986; R. Trümpy pers. comm. 1986)

and Kamm Beds).

Zone C. A zone with redeposited, commonly channelized sediments, derived from inner shelf zones A and B (Rankweil Ramp and proximal outer shelf; including the distal inner shelf in late Aptian time as well; Mittagspitz Formation, Klaus, and Rankweil Beds).

Zone D. A zone with moderate sediment accumulation rates, in which muddy and marly or calcareous, commonly laminated, hemipelagic sediments were deposited (outer shelf; max. 10 m/my; Mittagspitz Formation, Freschen and Hochkugel Beds, Seewen Formation).

The zonal distribution of Mid-Cretaceous helvetic shelf sediments is interpreted as the result of a bottom-hugging, westbound, geostrophic current system, the zone of ultralow sediment accumulation rates (zone B) being located within the zone of active erosion along the current axis. Eroded and/or win-

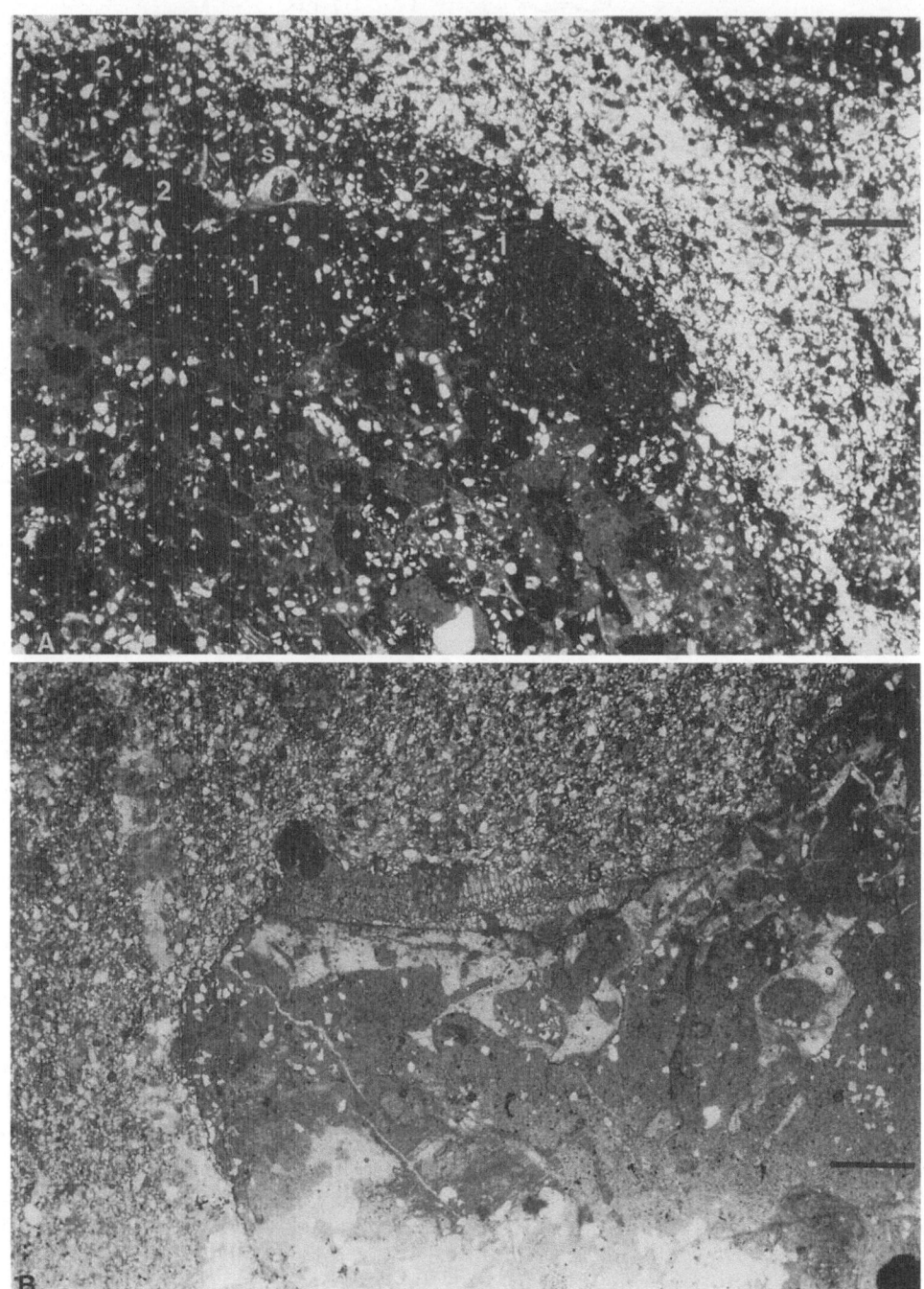

nowed particles were deflected seaward by smaller gradient currents and accumulated in the zone of redeposited sediments (zone C) and partly in the outer zone of hemipelagic sedimentation (zone D; e.g., Swift and Rice 1984).

Phosphatized fossil debris and lithoclasts within the phosphatic beds give further evidence for current activity. Autochtonous phosphatic crusts and a majority of the phosphatized particles in the condensed and winnowed, and allochthonous phosphatic beds display more than one phosphate generation. In general, the accretionary phosphatic laminae and envelopes include different quantities and grain sizes of siliciclastic detritus. Generation surfaces may be coated with iron oxyhydroxides or are overgrown with encrusting microbial colonies, sessile foraminifera, bryozoans, and serpulids (Figs. 45 and 46). Occasionally, the phosphatic particles show truncation surfaces, which may be colonized by microbial mats (Fig. 46). The phosphatized fossils may include different types of noncoeval sediment infills (Fig. 47). Isolated phosphatized fossil molds may occur (Fig. 46). Within one phosphate bed, phosphatized guide fossils of different zones, as well as endo- and epibiontic fossils are intimately mixed.

The above features point to a complex evolution of the phosphatic beds including repetitive **"Baturin"** cycles of burial, phosphatization, reworking, and reexposure at the sea floor. Baturin cycling was the result of episodic, current-induced, winnowing and erosion events, which interrupted sediment accumulation processes (Baturin 1971; Kennedy and Garrison 1975; Birch 1979; Krajewski 1984; Riggs 1984; Mullins and Rasch 1985; Brandt 1985; O'Brien 1986; Carbone et al. 1987; Sect. 3.6).

Some phosphatic particles also show signs of transportation (e.g., Pedley and Bennett 1985; Föllmi et al. in press). Phosphatized fossils may display different stages of abrasion within one bed. Some nodular phosphatic beds bear older phosphatized ammonoids than the underlying, nonphosphatized beds (e.g., section 7 in Fig. 19).

Evidence for the westward flow direction of the current system is provided by the geometry and internal structures in the upper Lower to Middle Albian Sellamatt sandbody, which ends in a wedge shape toward the east. This geometry suggests the presence of "sandwiching", bifurcated, currents, the bifurcation point being located in the eastern part of the Vorarlberg inner shelf

Fig. 45. Thin-section photomicrographs of phosphatic particles within the Plattenwald Bed (bar = 1mm).
A. Phosphatized calcareous lithoclast (biopelsparite) displaying two accreted phosphatic laminae at the top (1 and 2). The interface of the two phosphate generations is encrusted by serpulids (s), indicating particle exhumation and reexposure at the sea bottom after phosphatization. The particle is embedded in a fine-grained glauconite sandstone (type A in Fig. 24).
B. Phosphatized lithoclast (derived from the Schrattenkalk Formation) encrusted by a nonphosphatized bryozoan colony (b). The particle is embedded in a fine-grained glauconite sandstone (type A in Fig. 24)

(Figs. 19 and 49; Sect. 2.6.1). Such a configuration is only possible with west-ward current flow directions. This direction is also suggested by the discovery of a single, angular, exotic gneiss cobble in Upper Albian beds of the Gar-schella Formation. The cobble has a Precambrian age of isotopic closure (^{87}Rb/^{86}Sr dating: 827.7 +/- 17 my). The closest *in situ* occurrence of granitic rocks, similar in mineral composition and age, is situated in the eastern part of the Bohemian Massif (in the Moldadanubicum, near Brno; R. Steiger pers. comm. 1984). This cobble was probably transported in the roots of a tree, drifting along with the westbound current (Fig. 48).

The position of the westbound northern tethyan current system was not fi-xed. It changed laterally through time, inducing shifts in the location of the zone of ultralow sediment accumulation rates (zone B), as well as of the zone of redeposited sediments (zone C; Figs. 49 and 57; Pinet and Popenoe 1985; Schumann and Li Van Heerden 1988). Major landward shifts occurred in the early Aptian, latest Aptian, and latest Albian. The early Aptian and the latest Albian shifts appear to have been slow, and were probably related to eustatic sea level changes (Sect. 3.8; Haq et al. 1987). The prominent latest Aptian shift was very rapid and accompanied by major erosion; this episode is related to a short phase of reinforced differential subsidence, probably induced by a tecto-nic event (Sects. 3.2.2 and 3.7.3). The shift does not correlate with a eustatic sea level change on the chart of Haq et al. (1987; Sect. 3.7.1).

The Albian is characterized by slow and small shifts of an apparently paired current system. Separation into two main current systems is indicated by the repetition of zones A and B within the inner shelf (zones A, B, A, B, C, D from the proximal inner shelf to the outer shelf; Ouwehand 1987), as well as by the geometry and internal structures of the Niederi and Sellamatt sandbodies. Current bifurcation may have been caused by the broadening and shallowing of the inner shelf toward the west, as well as to irregularities in sea bottom topography of the eastern Vorarlberg inner shelf, indicated by local facies anomalies (not described here; for details cf. Föllmi 1986).

The slow Albian current shifts gave rise to the formation of inversely gra-ded sequences, consisting of upward coarsening glauconitic sandstones, topped by particulate phosphatic beds (e.g., Durschlägi or Plattenwald Bed on top of upward coarsening Niederi Beds; Kamm Bed on top of upward coarsening Aubrig Beds; cf. Föllmi et al. 1988).

Fig. 46. Plattenwald Bed.
A. Phosphatized internal cast of *Hypacanthoplites* sp. On the left side, a mold of a smaller sized *Hypacanthoplites* is attached.
B. Reverse side of the above shown compound cobble. Its surface is truncated and encrusted by phosphatized microbial mats. This surface is interpreted as an erosional surface, reflecting reexposure of the fossil internal cast and the attached fossil negative, after a first early diagenetic phosphatization phase. Microbial mats covered the erosional surface and were phosphatized in a se-cond phosphatization phase

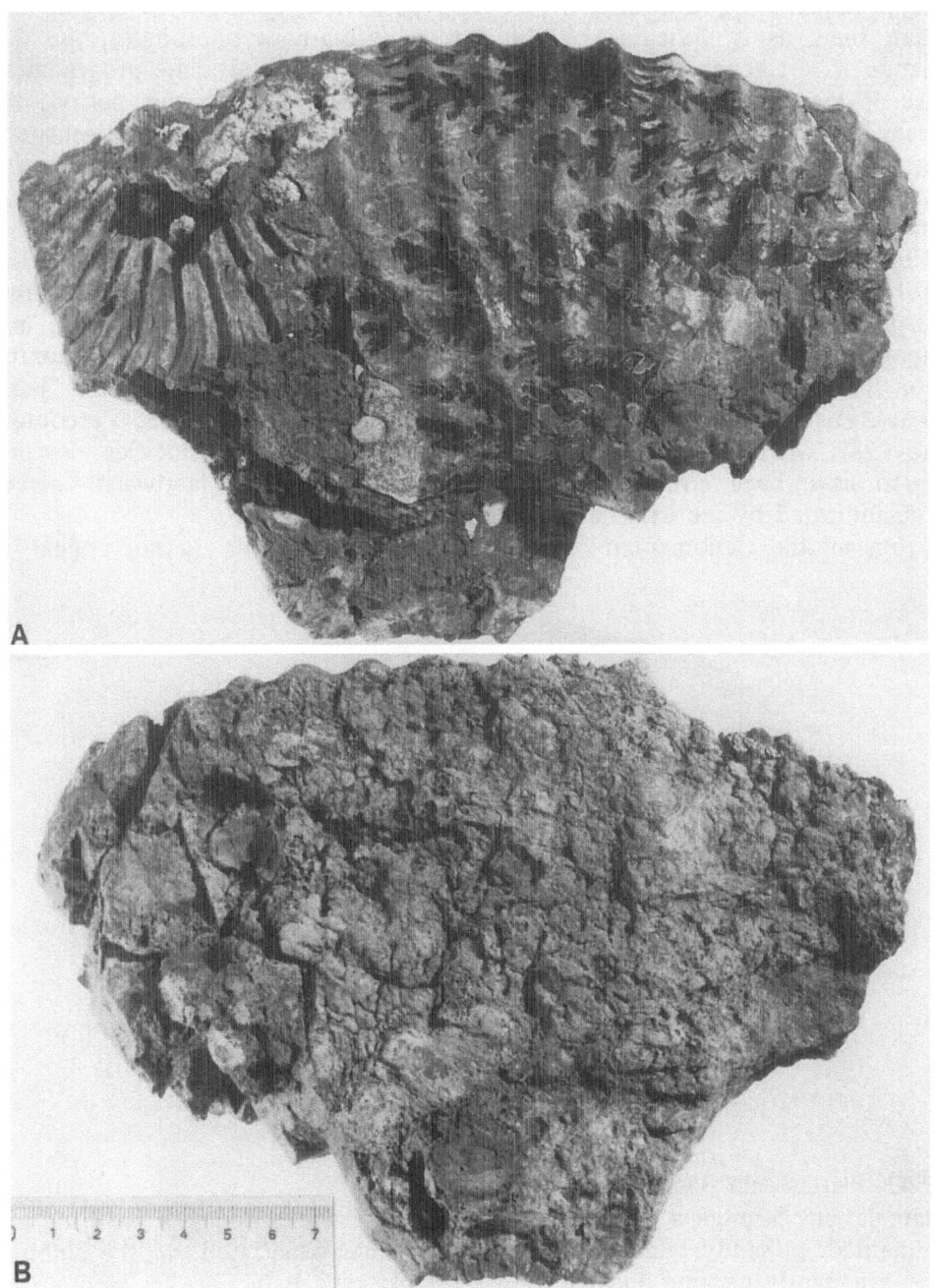

A

B

A prominent seaward shift of the current system occurred in middle late Aptian time, as is indicated by the lack of prominent phosphates, the dominance of terrigenous influenced sedimentation, and an ultimate progradation phase of the carbonate platform (Fig. 56; Sect. 3.8). During this period, the current system probably acted as a contouring outer shelf current, resuspending and redistributing fine-grained, hemipelagic, and turbiditic sediments (Sect. 2.4.1; Figs. 10 and 57). This seaward shift may have been related to an eustatic sea level fall (Sect. 3.7.1; Haq et al. 1987).

Current velocities of the inner shelf-based current system were in the order of 0.1-0.5 m/s (Ouwehand 1987). Current velocities may have been variable, in order to maintain a dynamic interplay between erosion, winnowing, and accumulation (Baturin cycling). This may have been induced by daily, monthly, or seasonal variations in tidal and wind energy, resulting in water "jets", "bursts", or "storms" (e.g., Kennett 1982; Reimers and Suess 1983; Vercoutere et al. 1987; Gross et al. 1988; Culvin et al. 1988). Current velocities also appear to have been stronger during or immediately after landward current shifts, indicated by increased amounts of eroded sediments.

From middle Cenomanian onward, preserved sediments do not appear to

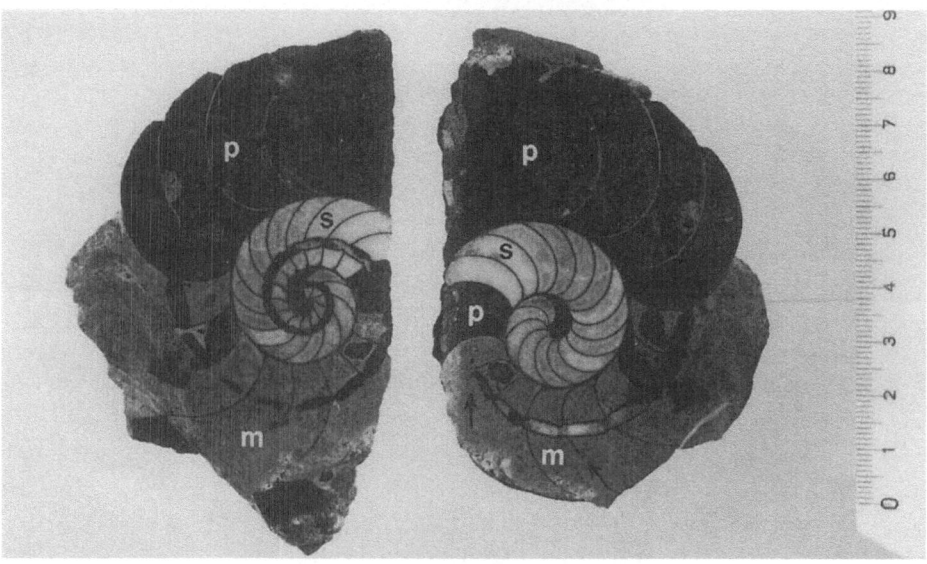

Fig. 47. Plattenwald Bed.
Equatorial cut through a nautiloid, in which different episodes of infilling are documented. p; sandy, glauconitic micrites, which penetrated *via* the siphone and subsequently became phosphatized (possibly type B in Fig. 24; *mammillatum* Zone or middle Albian). m; micrites comparable to Seewen micrites, which were infilled *via* defects in the phragmocone [note broken septa, (arrows); type H in Fig. 24; *dispar* Zone]. s; diagenetic sparry calcite.
The different infillings are indicative for a phase of reexposure at the sea floor and subsequent erosion of the shell after initial, early diagenetic phosphatization

Fig. 48. Angular gneiss cobble (oligoclase-microcline gneiss), embedded in muddy glauconitic sands. Interpreted as a dropstone, initially captured in a drifting tree

have been influenced by current activity. It is not clear whether this is due to diminishing current strength or to a further landward shift during Cenomanian time, beyond the exposed sediments of the helvetic shelf.

3.5 Oxygen Minimum Zone

Middle Upper Aptian to Upper Albian Freschen Beds in the outer shelf zone D display laminations, are weakly bioturbated, and generally lack autochthonous benthic organisms. Planktonic foraminifera (e.g., *Hedbergella*) are commonly internally phosphatized or have pyritized shells. Apparently, these hemipelagic sediments accumulated below poorly oxygenated bottom waters. Episodic turbidity currents introduced pioneering infauna and short increases in oxygen levels, documented by the presence of distinct bioturbation in intercalated turbidites and subjacent sediments (Fig. 13).

In contrast, inner shelf glauconitic sands and marls (zone A) are generally well-bioturbated and indicate the presence of better oxygenated bottom waters at that time. Therefore, the area of ultralow sediment accumulation rates and prominent phosphatization (zone B) appeared to have coincided with the upper boundary zone of a present oxygen minimum zone (OMZ). This observation is consistent with the notion that phosphogenesis favorably occurs and occurred near OMZ edges (Burnett 1977, 1980; Burnett and Roe 1983; Burnett et al. 1980, 1982; Arthur and Jenkyns 1981; Pisciotto and Garrison 1981; Mullins and Rasch 1985; Slansky 1986; Föllmi 1988).

Phosphatized fossils preserved in zone B suggest a food web dominated by

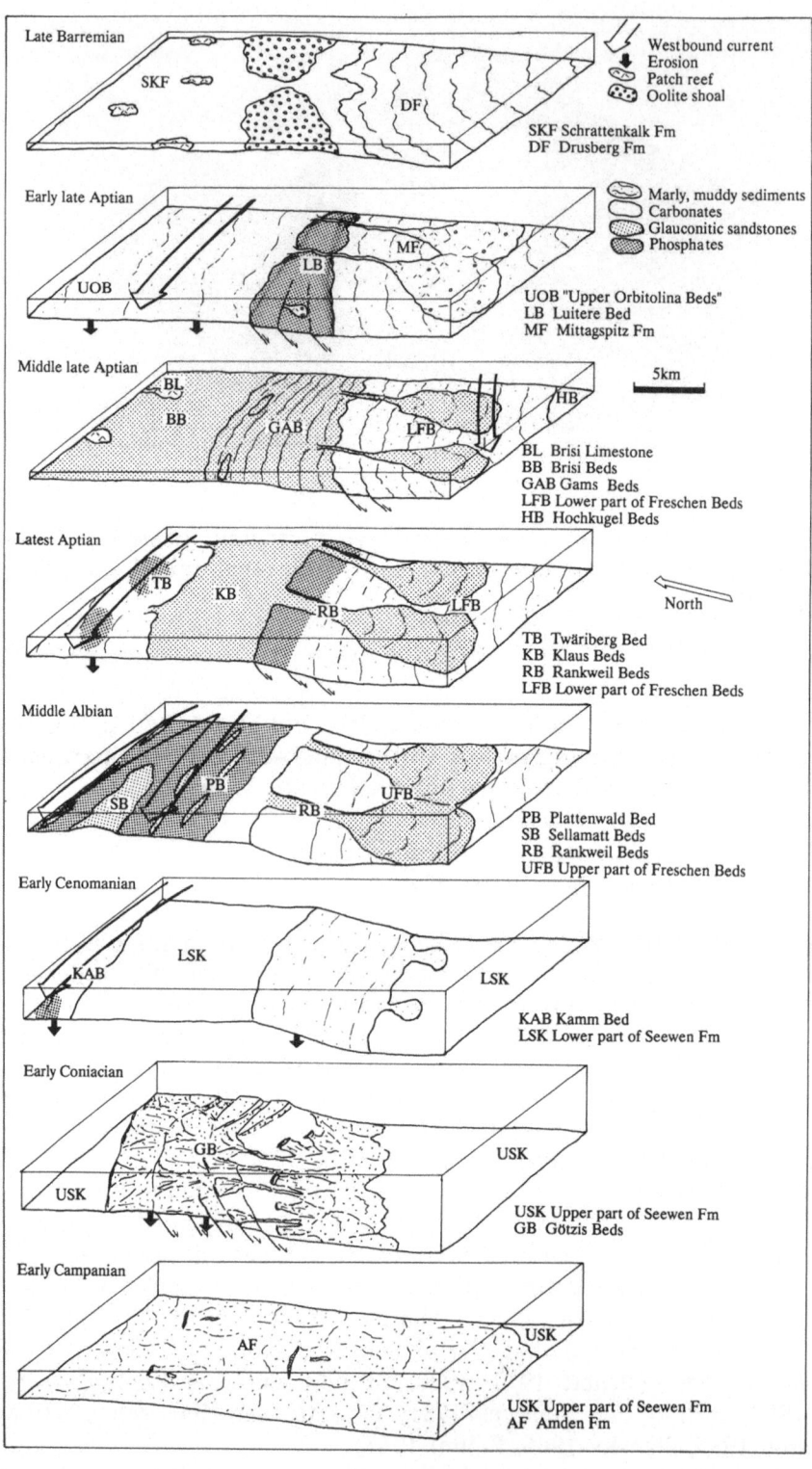

Late Barremian

SKF

DF

Westbound current
Erosion
Patch reef
Oolite shoal

SKF Schrattenkalk Fm
DF Drusberg Fm

Early late Aptian

UOB

LB

MF

Marly, muddy sediments
Carbonates
Glauconitic sandstones
Phosphates

UOB "Upper Orbitolina Beds"
LB Luitere Bed
MF Mittagspitz Fm

Middle late Aptian

BL
BB
GAB
LFB
HB

5km

BL Brisi Limestone
BB Brisi Beds
GAB Gams Beds
LFB Lower part of Freschen Beds
HB Hochkugel Beds

Latest Aptian

TB
KB
RB
LFB

North

TB Twäriberg Bed
KB Klaus Beds
RB Rankweil Beds
LFB Lower part of Freschen Beds

Middle Albian

PB
SB
RB
UFB

PB Plattenwald Bed
SB Sellamatt Beds
RB Rankweil Beds
UFB Upper part of Freschen Beds

Early Cenomanian

LSK
KAB
LSK

KAB Kamm Bed
LSK Lower part of Seewen Fm

Early Coniacian

GB
USK
USK

USK Upper part of Seewen Fm
GB Götzis Beds

Early Campanian

AF
USK

USK Upper part of Seewen Fm
AF Amden Fm

prolific microbial communities (Figs. 26, 27, 46, 51, 52, and 53), gastropods feeding on the microbial mats, suspension feeders such as solitary corals, sponges, brachiopods, bivalves, and crinoids, as well as predators such as naticid gastropods, ammonoids, and sharks. Endobiontic bivalves and echinoderms appeared as well. The rich and diverse fauna, preserved in zone B, represents a typical eutrophic community, proliferating in probably nutrient-enriched bottom waters along the upper OMZ boundary (Thompson et al. 1985; Vercoutere et al. 1987). The shelly epifauna is especially abundant in proximal parts of zone B, indicating sufficient oxygen levels ("mixed layer"), whereas microbial communities generally thrived in distal parts of zone B, probably indicating dysaerobic conditions (Fig. 51; Williams and Reimers 1983; Reimers et al. in press).

The microbial mats possibly formed an important nearby source of particulate organic matter, because of their susceptibility to disruption and suspension by current activity (Grant and Bathmann 1987).

Positions of the OMZ upper boundary zone and the current system probably were interdependent, the OMZ being limited in its upward extension by the current system, in an analogous fashion to the northbound undercurrent along the upper OMZ boundary zone offshore central California (Vercoutere et al. 1987).

Lateral shifts in the current system may have induced vertical fluctuations of the upper OMZ boundary zone, and thus lateral shifts of the position of zone B (Figs. 6, 10, 16, 20, 30, 49, and 51).

Upwelling features are absent in Aptian-Lower Cenomanian helvetic sediments. A short food chain with abundant opportunistic, siliceous phytoplankton and fish, typical for upwelling areas, is unknown from the helvetic shelf (e.g., Hallock 1987). Biogenic siliceous sediments are accordingly not very common (e.g., Parrish 1982; Parrish and Curtis 1982; Iijima et al. 1983). The occurrence of an OMZ was rather related to a general and widespread oxygen deficiency in shelf and basinal bottom waters during late Aptian and Albian times ("anoxic event"; e.g., Schlanger and Jenkyns 1976; Arthur and Schlanger 1979; Bralower and Thierstein 1984, 1987; Summerhayes 1987).

3.6 Aptian to Early Cenomanian Phospho- and Glaucogenesis

Phosphogenesis occurred almost continuously during a period of approximately 23 my, from earliest Aptian to early Cenomanian, with a short decline during the middle late Aptian time interval (Luitere, Twäriberg, Durschlägi, Wannenalp, Plattenwald, and Kamm Beds).

The main site of phosphogenesis was located in zone B, along the erosive current axis/axes and near the OMZ upper boundary level (Fig. 51). Minor

Fig. 49. Schematic overview of the evolution of the Vorarlberg helvetic shelf. Left = inner shelf; right = outer shelf. Vertical exaggeration is 10:1

phosphogenesis took also place outside this zone (e.g., facies D of the Luitere Bed, Freschen Beds, Brisi Beds).

The Mid-Cretaceous helvetic phosphatic beds fall into three major groups (Fig. 50; Föllmi et al. in press):

1. **Pristine** (= nonreworked) **phosphates,** including penetratively phosphatized, rugged surfaces of previously lithified sediments, commonly covered by well-preserved phosphatized microbial mats. The presence of surficially phosphatized, subvertical fissures and angular lithoclasts suggests that lithification occurred prior to phosphatization (e.g., Luitere Bed on top of the Schrattenkalk Formation; Twäriberg Bed on top of the Brisi Limestone; Kamm Bed on top of Aubrig Beds).
Phosphatization of previously "soft", nonlithified sediments also belong to this category of phosphates. These sediments commonly include phosphatized "microstromatolites", i.e., microbial colonies, which acted as sediment stabilizers (e.g., Krajewski 1984). These phosphatized sediments are occasionally topped by phosphatized microbial mats (e.g., Plattenwald Bed; Fig. 26).
2. **Condensed phosphates,** consisting of an assemblage of phosphatized, angular lithoclasts and phosphatized, commonly well-preserved fossils, embedded in slightly to distinctly younger sediments. Characteristic of multi-event winnowed phosphatic beds, generated over longe time periods, is the presence of different phosphate generations, and an intimate mixture of fossils of different time zones and habitats within a bed.
In the distal part of zone B, characterized by low oxygen levels in its bottom waters, proliferating microbial mats, and strong condensation, stacks of autochthonous (formerly pristine) phosphatic crusts formed (Fig. 50).
3. **Allochthonous phosphates,** present within mixtures of commonly rounded phosphatic and nonphosphatic particles. This type of phosphate appears in gravity flows, or as fine-grained admixtures in winnowed, condensed beds.

These three groups are end members. Lateral and vertical transitions, as well as hybrids, are observed. For example, condensed and winnowed phosphatic beds usually include allochthonous particles, possibly resulting from varying energy conditions and/or varying particle sizes; condensed or allochthonous phosphatic particles can be cemented by autochthonous phosphatic sediments; pristine phosphates can be superposed by winnowed beds or *vice versa* (Fig. 50; Föllmi et al. in press).

3.6.1 Low Sedimentation Rates, Microbial Mats, Organic Matter, PO_4^{3-} Flux, and Phosphogenesis

As mentioned above, microbial mats and colonies were abundant throughout the zone of ultralow sediment accumulation rates and major phosphogenesis (zone B). Their close relationship with phosphates and hemipelagic micrites suggests a "deep-water" character ("deep-water stromatolites", e.g., Krumbein

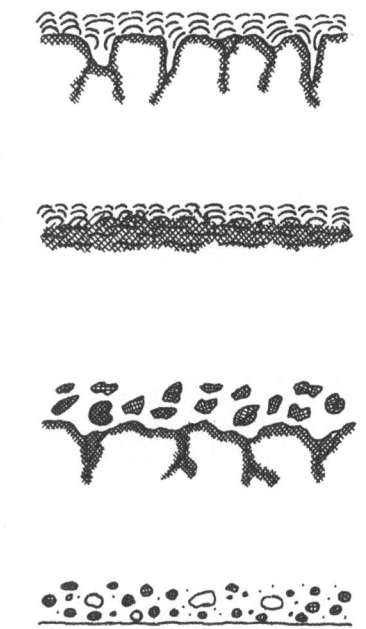

PRISTINE PHOSPHATIC BEDS

Peripheral phosphatization of
lithified sediments;
cover of phosphatized microbial mats

Phosphatization of unlithified
sediments, interspersed with
microbial colonies ("micro-stromatolites");
cover of phosphatized microbial mats

CONDENSED PHOSPHATIC BEDS

Phosphatized fossil debris and lithoclasts;
fossils are well preserved, clasts are angular

ALLOCHTHONOUS PHOSPHATIC BEDS

Phosphatized particles, mixed with non
phosphatized lithoclasts. Particles and lithoclasts are abraded

Fig. 50. Types of stratification in the helvetic Mid-Cretaceous phosphatic beds. Thicknesses of the sequences are generally <50 cm. These classes depict end members, which enclose transitional ranges. Transitions are possible both in parallel and perpendicular directions to depositional strike, and in time

1983; Williams 1984; Delamette et al. 1984).

Phosphatized microbial mats are a major constituent of pristine and condensed laminated phosphatic beds (Fig. 50), and appear as well in condensed particulate and allochthonous beds, in the form of thin envelopes around particles ("coated grains"), or of mm-thin rims within fossil debris (Figs. 26, 27, 52, and 53; Krajewski 1984; Föllmi 1986; Ouwehand 1987; Delamette 1988a).

The intimate association of microbial mats and colonies, and phosphates within the Garschella Formation is consistent with the general notion that microbial activity and phosphogenesis are closely related processes (e.g., Cayeux 1936; O'Brien et al. 1981; Baturin 1982; Soudry and Champetier 1983; Krajewski 1984; Williams 1984; Lucas and Prévôt 1984, 1985; Dahanayake and Krumbein 1985; Soudry 1987; Lamboy and Monty 1987; Soudry and Lewy 1988; Reimers et al. in press).

The Garschella microbial mats generally show an excellent textural preservation (Figs. 26 and 52; Delamette 1981, 1988a; Delamette et al. 1984; Föllmi

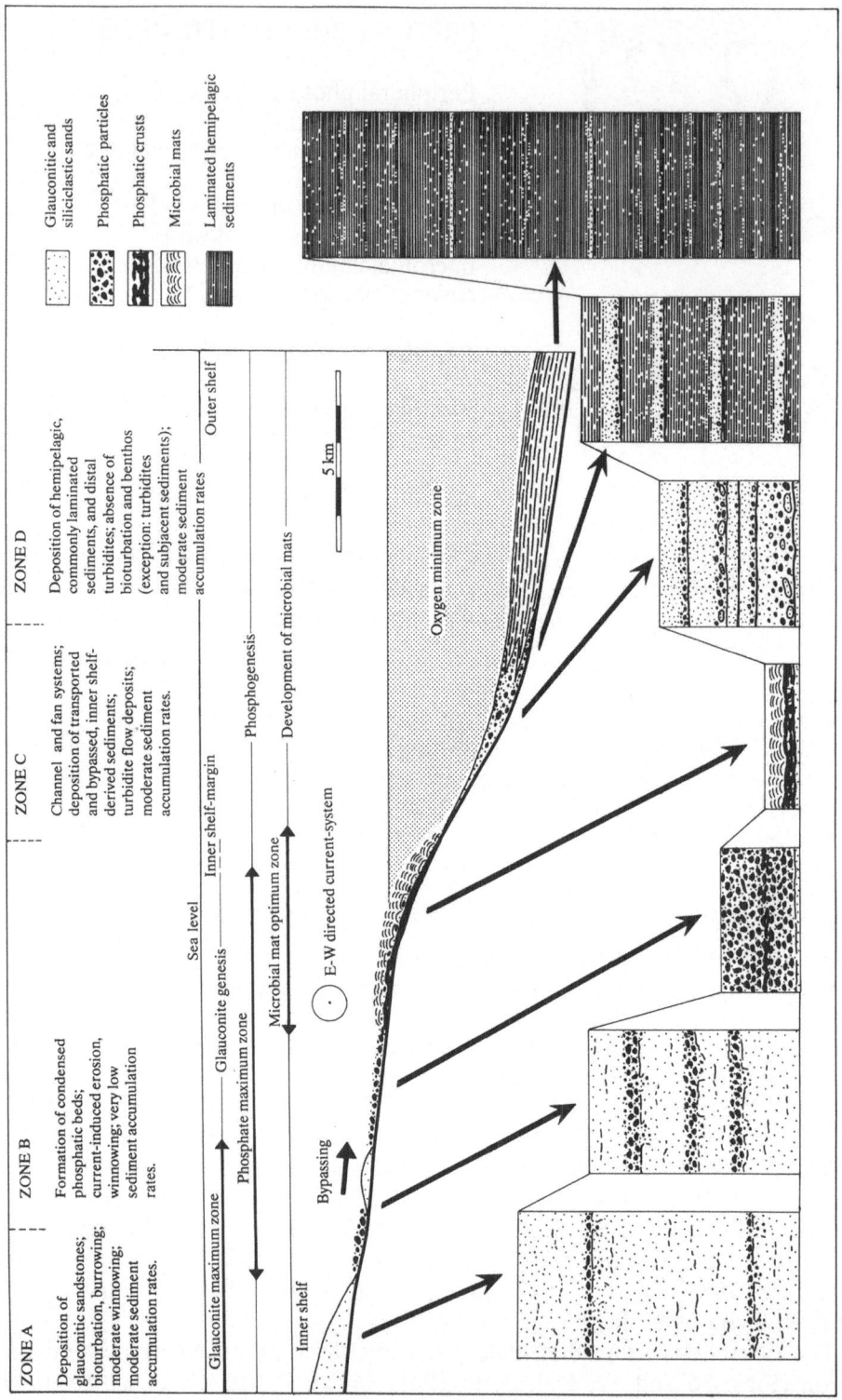

1986; Ouwehand 1987). This is suggestive of phosphatization of the mats prior to decay of the microbial organic matter. It appears as if the microbial mats were "frozen" by the process of phosphatization (Krajewski 1984). The presence of these early-stage phosphates indicates that phosphogenesis, although strongly linked to the presence of microbial colonies, was not exclusively dependent on the PO_4^{3-} flux of decaying microbial organic matter within the uppermost sediments (e.g., Broecker and Peng 1982; Froelich et al. 1983; Jahnke et al. 1983; Ouwehand 1986, 1987; Reimers et al. in press).

This observation brings up the general question of how downward concentration gradients of dissolved PO_4^{3-} into the sediments could have been initiated and maintained, in order to accomodate phosphogenesis and the observed amounts of phosphates, taking into account the ultralow sediment accumulation rates during the formation of the helvetic phosphatic beds.

The early diagenetic alteration of PO_4^{3-}-rich organic matter was probably of subordinate importance (Burnett 1977; Suess 1981; Ouwehand 1986, 1987), since phosphatization of microbial organisms occurred in an early stage, probably prior to the decay of its organic matter, and the sediment accumulation rates in zone B were too low to bury sufficient organic matter, even when the effect of Baturin cycling is considered. Ouwehand (1986, 1987) showed in a quantitative approach that, assuming a PO_4^{3-} flux entirely dependent on early diagenetic alteration of organic matter and assuming inclusion rates of 2% reactive organic matter into sediments, a total sediment accumulation in the range of 35-100 m is needed to accumulate 7-15 cm of apatite. Actually, such sediment accumulation rates were never achieved in zone B. Sediment accumulation rates in the phosphatic beds varied between 2-20 cm/my. Even when the sediment accumulation rates of redeposited sediments beyond zone B are added (zone C; Klaus and Rankweil Beds; max. rates are 1 m/my), bulk sediment accumulation rates did not exceed 1 m/my.

In addition, Baturin cycling may actually have reduced the quantities of organic matter within the sediments, because of frequent resuspension and subsequent oxidation of organic matter within bottom waters (Reimers & Suess 1983).

Additional factors minimizing inclusion rates of organic matter into the sediments are probably represented by the abundant microbial mats, which re-

Fig. 51. Diagrammatic transect perpendicular to the shelf, displaying the position of zones A-D in a setting where the current system is located in the distal inner shelf (e.g., Plattenwald Bed). Sections in lower half of figure are representative and approximately time-equivalent. Note the areas of optimal phosphogenesis ("phosphate maximum zone"), glaucogenesis ("glauconite maximum zone") and prolifering growth of microbial mats ("microbial mat optimum zone"). Glaucogenesis is limited on the inner shelf, whereas phosphogenesis occurred throughout the helvetic shelf. Current-induced Baturin cycling is limited to zone C, and is directly responsible for physical concentration of phosphates

cycle organic matter, and by the high degree of other biological activities such as suspension feeders and endobenthic fauna, consuming organic matter (e.g., Müller and Suess 1979; Berner 1982; Reimers et al. in press). It is therefore conceivable that, notwithstanding the possible presence of elevated particulate organic matter levels in the bottom waters of zone B, burial rates of organic matter into the sediment were minimal, due to current activity, very low to zero sediment accumulation rates, and the presence of microbial mats and a specialized fauna at or near the sediment-water interface.

The only way to bring major quantities of organic matter into the sediment was by sudden, catastrophic burial of entire biotic communities. This process probably took place in zone B and it probably catalyzed phosphogenesis (see below). But then again the amount of enclosed organic matter would not have been sufficient to account for the observed amounts of apatite (within one generation; e.g., Ouwehand 1986, 1987).

It seems necessary for these reasons to invoke a second mechanism of PO_4^{3-} concentration in the process of phosphogenesis, decoupled from the early diagenetic alteration of buried PO_4^{3-}-rich organic matter (Froelich 1984; Froelich et al. 1988; O'Brien and Heggie 1988; Heggie et al. in press; Delaney and Boyle 1988; Kastner pers. comm. 1988; Garrison pers. comm. 1988). This mechanism most probably supplied dissolved PO_4^{3-} from bottom waters into the sediment. The bottom waters within the OMZ upper boundary zone probably included elevated amounts of dissolved PO_4^{3-}, due to longer residence times of organic matter (induced by current activity and Baturin cycling), which is therefore more subjected to decomposition (e.g., Soutar et al. 1981; Baturin 1982; Kennett 1982; Reimers and Suess 1983; Vercoutere et al. 1987).

The concentration of dissolved PO_4^{3-} into interstitial waters may have taken place via physicochemical cycles of iron (and probably manganese) oxyhydroxides in the way Shaffer (1986), O'Brien and Heggie (1988), Heggie et al. (in press), and Froelich et al. (1988) described (cf. Stumm and Leckie 1970; Manheim et al. 1975; Filipek and Owen 1981; Jansson 1986). Iron and manganese cycle between oxic and anoxic zones, taking up PO_4^{3-} from bottom waters in oxidized, solid states, diffusing downward below the sediment-water interface, releasing PO_4^{3-} in reduced states near the oxic-anoxic interface or in local anoxic microenvironments (e.g., near decaying organic matter), diffusing upward into the oxic zone in dissolved states, taking up PO_4^{3-} again, etc. ("phosphate pump and shuttle" of Shaffer 1986). The release of PO_4^{3-} during reduction of the iron and manganese oxyhydroxides is probably mediated by microbial activity (Jansson 1986).

Fig. 52. Phosphatized microbial mats and colonies.
A. Phosphatized columnar microbial mats encrusting surficially phosphatized Schrattenkalk pebbles. Note tilted microbial mat on left pebble, overgrown by normally oriented microbial mats.
B. Thin-section photomicrograph of microbial mat in A. (bar = 1 mm). Note the presence of encrusting agglutinated foraminifera (Placopsilina sp.).

3.6.2 Catastrophic Burials, Inner Shelf Equilibria, and Phosphogenesis

Phosphogenesis in zone B, the zone of very low sediment accumulation rates, occurred under dynamic paleoceanographic conditions, which were induced by the westbound current system (Fig. 51; Sect. 3.4). The vital importance of this current system to Mid-Cretaceous helvetic phosphogenesis is demonstrated by the Albian phosphatic system (Sect. 2.6.1).

In the Albian, three phases of sand replenishment are observed, which pushed the boundary of the proximal extension of zone B in a seaward direction (Niederi Beds; boundary Lower/Upper *tardefurcata* Zone; Sellamatt Beds; Upper *mammillatum* zone; and Aubrig Beds; Upper *inflatum* Zone; Fig. 2).

The youngest sediments with dominant siliciclastic fractions included in the Plattenwald Bed (representing zone B) suggest that (Fig. 24)

1. Distal areas within zone B were not influenced by these phases of sand replenishment. Siliciclastic sands, once present in these areas, were probably derived from middle Upper Aptian Brisi and Gams Beds: they were completely removed at the early/late *tardefurcata* Zone boundary. Since then, these areas have been in equilibrium (Hamilton et al. 1980; Dott, Jr 1988; Smith 1988).
2. Intermediate areas of zone B were influenced by the oldest sand replenishment phase and reached equilibrium at the end of the *mammillatum* Zone.
3. More proximal areas of zone B have been influenced by the oldest and middle sand replenishment phases and were sand depleted in the *loricatus* Zone.
4. The most proximal areas of zone B were affected by all three sand replenishment phases and reached only partial equilibrium (Fig. 19).

These observations suggest that the area of sand depletion within zone B sequentially prograded landward during Albian time, retarded by phases of sand replenishment. This was probably the result of the combination of westbound current system activity and a slow general eustatic sea level rise (Sects. 3.4 and 3.7.1). However, during this period, zone C, zone of redeposition beyond zone B (Sect. 3.4), was not excluded from receiving siliciclastic detritus. The Rankweil Beds contain sediments, rich in siliciclasts, of uppermost Aptian to uppermost Albian age (Sect. 2.6.1). This strongly suggests that zone B not only served as a source of winnowed and eroded particles, included in zone C, but also as a <u>bypass area</u> for siliciclastic detritus that directly was transported from zone A to C.

The siliciclastic sands may have bypassed zone B in two different ways (e.g., Hamilton et al. 1980; Flemming 1988; Smith 1988):

1. In the form of <u>palimpsest sandbodies</u>. Such sandbodies are adjusted to bedforms with maximum stability in a stable current system with constant velocities (e.g., Stride 1988). Changes in current velocities and in the position of the current axis would reactivate and dislocate such bodies. At the

moment of complete removal and inclusion of these sand packages into zone C, equilibrium is reached.

2. In the form of <u>bedload transport</u>. The transfer of sand from zone A to zone C *via* shelf zones that already reached equilibrium may have occurred exclusively in this manner.

Important in this context is the notion that phosphogenesis in zone B apparently was related to the presence of siliciclastic sediments (Sect. 2.6.1). As soon as an area within zone B reached equilibrium, phosphogenesis ceased, although current-induced low sediment accumulation rates and condensation persisted ($T_{PH} < T_C$, Sect. 2.6.1).

A further observation, important to the reconstruction of phosphatization processes, is that of the already mentioned excellent textural preservation of the included phosphatized microbial mats. Other fossil groups display a good conservation as well. Sponges, which are usually prone to rapid postmortem decomposition, are generally preserved as body fossils; echinoderms, normally victim of early diagenetic calcite cementation, preserved an unaltered stereom; ammonoids may preserve fine ornamental structures. Good fossil preservation is generally known from phosphatic sediments (e.g., Kennedy and Garrison 1975; Baturin 1982; Ouwehand 1987; Soudry and Lewy 1988). The presence of

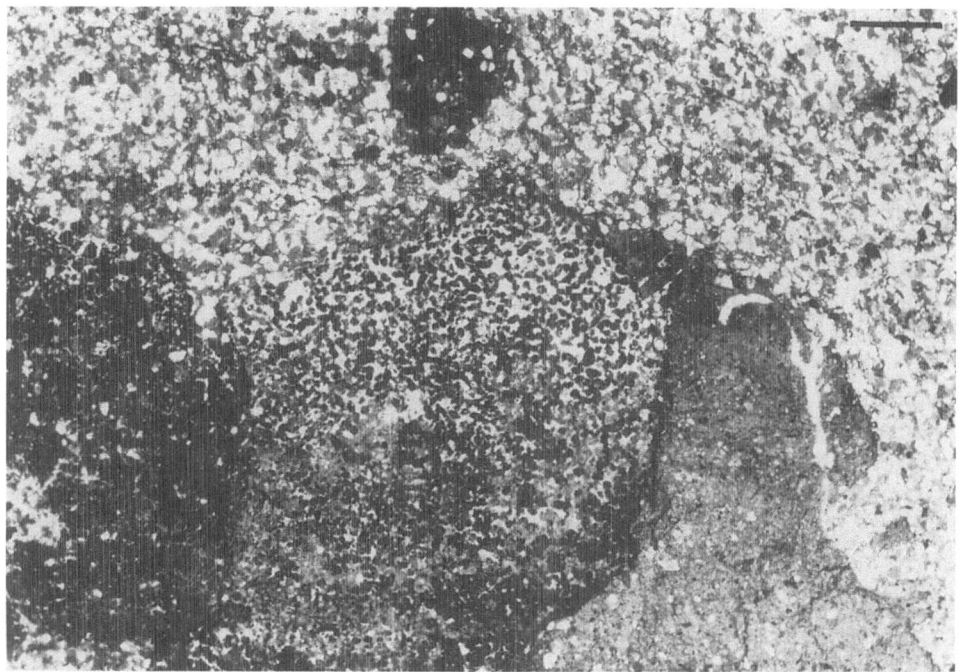

Fig. 53. Plattenwald Bed; thin-section photomicrograph (<u>bar</u> = 1 mm). Phosphatized rod-like corpuscules in an ammonoid chamber, probably of bacterial origin (e.g., Soudry and Champetier 1983; "cluster cement" in Krajewski 1984, Fig. 9, 13; Delamette 1985)

abraded, erosional surfaces on phosphatized fossils is due to reworking and transportation after phosphatization.

The state of preservation of these phosphatized fossils conjures up a "Pompeii"-like scenario, in which the organisms were victims of instantaneous burial. This probably was the case: catastrophic burials may have been caused by sudden dislocations of palimpsest sands, covering entire biotic communities under sediment ("obrution"; Seilacher et al. 1985; Speyer and Brett 1988; Brett and Seilacher in press; Fig. 54). Obrution-type burial is also suggested by the abundance of benthic and benthos-oriented nektonic organisms (ammonoids), and the rarity of genuine nektonic organisms (fish), preserved in zone B.

In the chaotic, metastable, and reactive environments of an obrution deposit, phosphogenesis may have taken place in a scenario similar to the following (Figs. 51 and 54).

1. The westbound current system causes very low to zero sediment accumulation rates. Bottom waters are probably enriched in nutrients, due to the proximity of the OMZ and nutrient recycling (Arthur and Jenkyns 1981; Mullins and Rasch 1985; Mullins et al. 1985; Hallock 1987). Sediments at the sea floor are generally stabilized by microbial mats and colonized by a rich community of benthic grazers and suspension feeders, and their predators.
2. Lateral shifts of the current system and changes in current velocity lead to perturbations within this "steady-state" system. Sediments are reactivated or resuspended, and sandbodies dislocated; biotic communities are disrupted and buried catastrophically under palimpsest siliciclastic sands.
3. The large amounts of organic matter, suddenly buried under sheets of sand, attract endobiontic, bioturbating, and scavenging organisms (e.g., indicated by phosphatized *Palaeophycus* trails).
4. Microbial activity around the buried organic matter leads to the development of local anoxic environments (e.g., around larger organisms), or induces a shallow oxic/anoxic interface (e.g., in the case of buried microbial mats).
5. The presence of a bioturbating endofauna enhances the permeability of the sandsheet, which in turn enhances microbial growth rates (e.g., Yingst and Rhoads 1980). Rapid sediment-mixing rates, induced by bioturbation, and bottom current activity promote the circulation of water in the sandsheet (irrigation). The high permeability and enhanced water circulation favor the physicochemical cycling of iron and manganese oxyhydroxides, and hence the concentration of PO_4^{3-} near oxic/anoxic boundaries within the sandsheet (e.g., Heggie et al. in press).

Fig. 54. A possible scenario for the development of multi-event condensed phosphatic beds. The cycle of catastrophic burial, phosphatization and reexposure is repetitive (Baturin cycling; indicated by thinner arrows). T is an estimation of the time span in which an event or phase took place

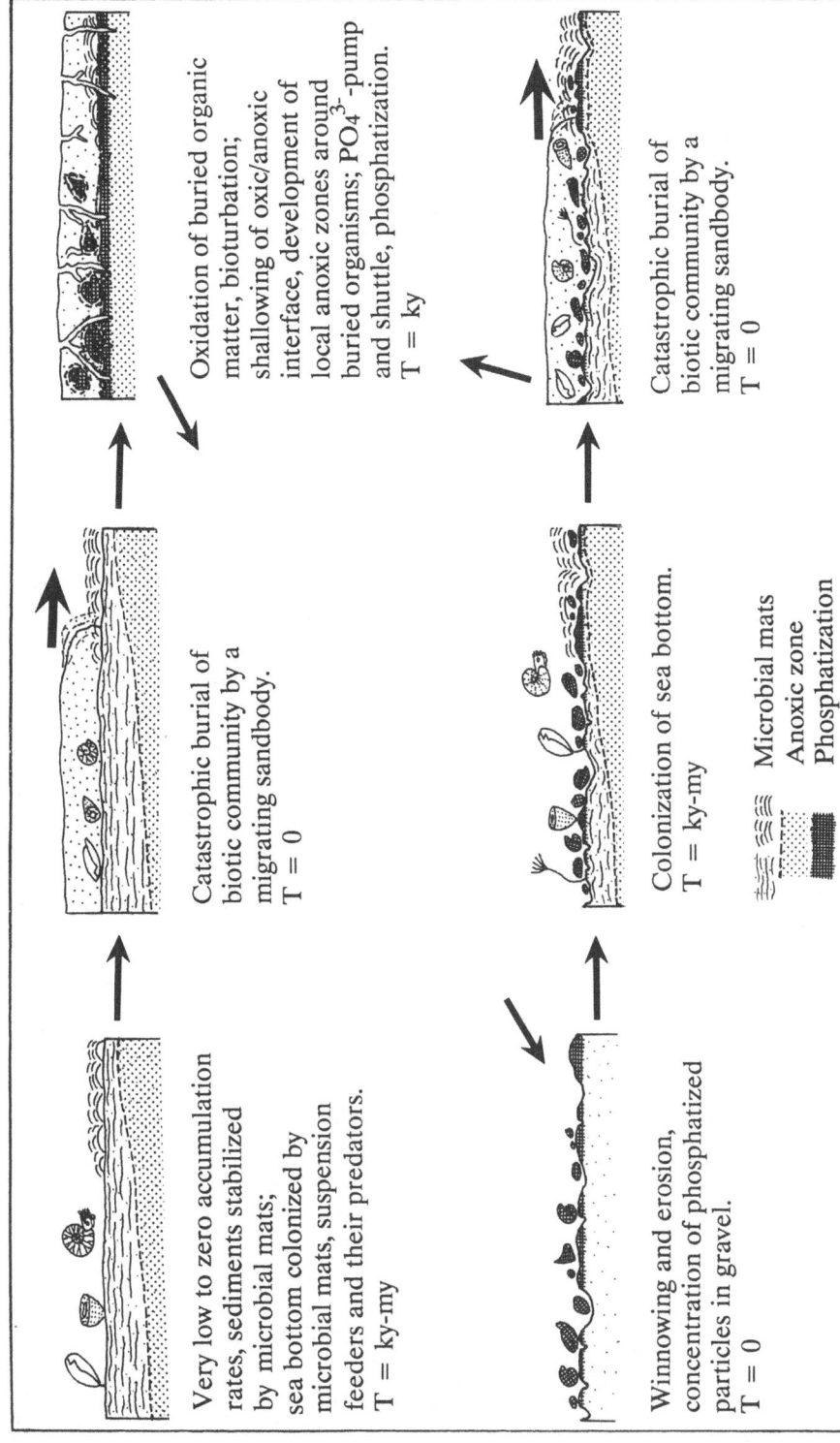

Very low to zero accumulation rates, sediments stabilized by microbial mats; sea bottom colonized by microbial mats, suspension feeders and their predators.
T = ky-my

Catastrophic burial of biotic community by a migrating sandbody.
T = 0

Oxidation of buried organic matter, bioturbation; shallowing of oxic/anoxic interface, development of local anoxic zones around buried organisms; PO_4^{3-}-pump and shuttle, phosphatization.
T = ky

Winnowing and erosion, concentration of phosphatized particles in gravel.
T = 0

Colonization of sea bottom.
T = ky-my

Catastrophic burial of biotic community by a migrating sandbody.
T = 0

Microbial mats
Anoxic zone
Phosphatization

6. The high PO_4^{3-} concentrations lead to precipitation of apatite near the oxic-anoxic interface (redox boundary) and within the local anoxic micro-environments. Apatite possibly nucleates around microbial phosphate-rich granules (e.g., Reimers et al. in press).

7. Variations in current velocity or lateral shifts in the current system cause winnowing and removal of the covering siliciclastic sands, until a gravel enriched in phosphatized particles is formed (start of Baturin cycling). Strong current impulses may also remove and transport the phosphatic particles (size and weight-dependent), and induce their inclusion in allochthonous phosphatic beds (e.g., in zone C; Fig. 51).

8. A steady-state situation is reintroduced, similar to 1, except for the presence of phosphatized particles, and cycle 2 to 7 repeats (Baturin cycles; Fig. 54).

Employing this scenario to the genesis and evolution of Mid-Cretaceous, ultra-condensed phosphatic sediments, explains (1) how larger amounts of relatively undegraded organic matter may be trapped and phosphatized in an environment of otherwise very low sediment accumulation rates (2-20 cm/my); (2) how organisms render their excellent fossil preservation in this environment ("Pompeii" scenario); (3) how Baturin cycles functioned; and (4) why the internal structure and composition of phosphatic beds differ so strongly laterally (Fig. 24).

The zone of prominent phosphogenesis and very low accumulation rates (zone B) is referred to as the "phosphate maximum zone" (Fig. 51; Föllmi 1988). Subordinate phosphogenesis occurred also in zones A and C and produced small, one-generation phosphatic particles (e.g., in Brisi and Freschen Beds). In this case, the sudden burial of organic matter may have played a decisive roll as well. However, major phosphogenesis and, particularly, phosphate enrichment has not been observed in the areas outside zone B, probably because of the presence of generally higher sediment accumulation rates and the absence of Baturin cycling.

3.6.3 Glaucogenesis

Glauconite is widespread in the inner shelf and Rankweil Ramp sediments of the Garschella Formation. It occurs in thin surficial coatings around micritized particles (e.g., in the Brisi Limestone), enclosed in phosphatized particles or on top of phosphatized crusts, in pore spaces of fossils (e.g., foraminifera, bryozoa, echinoderm fragments, or in commonly spherical particles (e.g., altered faecal pellets).

Glauconite particles are preserved *in situ*, or experienced transportation (detrital glauconites). Pristine (= nonreworked) glauconite particles are common in bioturbated marls, muds, and matrix-rich sandstones (e.g., basal Gams and Sellamatt Beds), and are distinguished by their abundance (locally forming "glauconitites"; i.e., glauconite contents over 50%), by their presence in non-sorted assemblages, and by their homogeneity in color and chemistry. Allochthonous particles appear in matrix-poor sandstones (e.g., Gams, Brisi, Niederi and Aubrig Beds), and are incorporated in the condensed phosphatic beds and in the redeposited sediments along the Rankweil Ramp. They are heterogeneous and well-sorted, adapting their grainsize to that of detrital siliciclastic particles.

Glauconite is most common in zone A, the detritus-rich inner shelf area (Fig. 51), both as autochthonous mineralization and allochthonous particles. Glauconite minerals of zone B, the zone of maximum condensation and phosphatization, are in many cases enveloped with a phosphatized rim ("glauco-apatites"; e.g., Haldimann 1977; Ouwehand 1987).

On the Vorarlberg shelf, glauconite formation occurred preferably in zone A, the inner shelf zone of moderate to high detritus-input, of moderate winnowing, of resulting net sediment accumulation rates of approximately 10-30 m/my, as well as of oxygenated bottom waters. In contrast, phosphogenesis favorably occurred in zone B, where the detrital influence was minimized, winnowing was prevalent and accumulation rates approached zero, and the bottom waters probably were oxygen-poorer (Fig. 51).

3.7 Relative Sea Level Changes, Unconformities, and Tectonic Events

3.7.1 Relative Sea Level Changes, and Their Probable Causes

Relative sea level changes in the helvetic realm resulted from an interplay between eustatic sea level changes, crustal subsidence, differential subsidence related to different sediment compaction rates in the inner and outer shelf, tectonic "events", and progradation of carbonate platforms and siliciclastic sandbodies. Current-induced erosion and shelf failure represent second-order mechanisms with a local, smaller effect on the relative sea level.

Inferences from facies analyses and interpretations of sedimentary structures within the helvetic sediments allow for a qualitative compilation of relative sea level changes (Fig. 55; left column; based on sediments of the Vorarlberg distal inner shelf; cf. Figs. 3 and 56). A definite determination of primary causes of each sea level change is difficult and only apparent correlations can be made.

A relative sea level rise in early and early late Aptian times is documented by the transition from the formation of Schrattenkalk shallow-water carbonates to the development of a condensed phosphatic bed, including pelagic micrites (Luitere Bed; Sects. 2.2, 2.3, and 3.8; Figs. 56 and 57). This slow and steady sea level change may have been induced by a combination of crustal subsidence (10-13 m/my; Funk 1985), current-induced erosion on the platform (1-10 m/my; Fig. 6), and a general eustatic sea level rise (Harris et al. 1984; Haq et al. 1987).

The relative sea level fall in middle late Aptian time is difficult to judge. Progradation of shallow-water carbonates (Brisi Limestone) and adjacent siliciclastic sands (Brisi Beds) may have been favored by the seaward shift of the *westbound* current system (Figs. 10 and 49; Sect. 3.4), and was a consequence

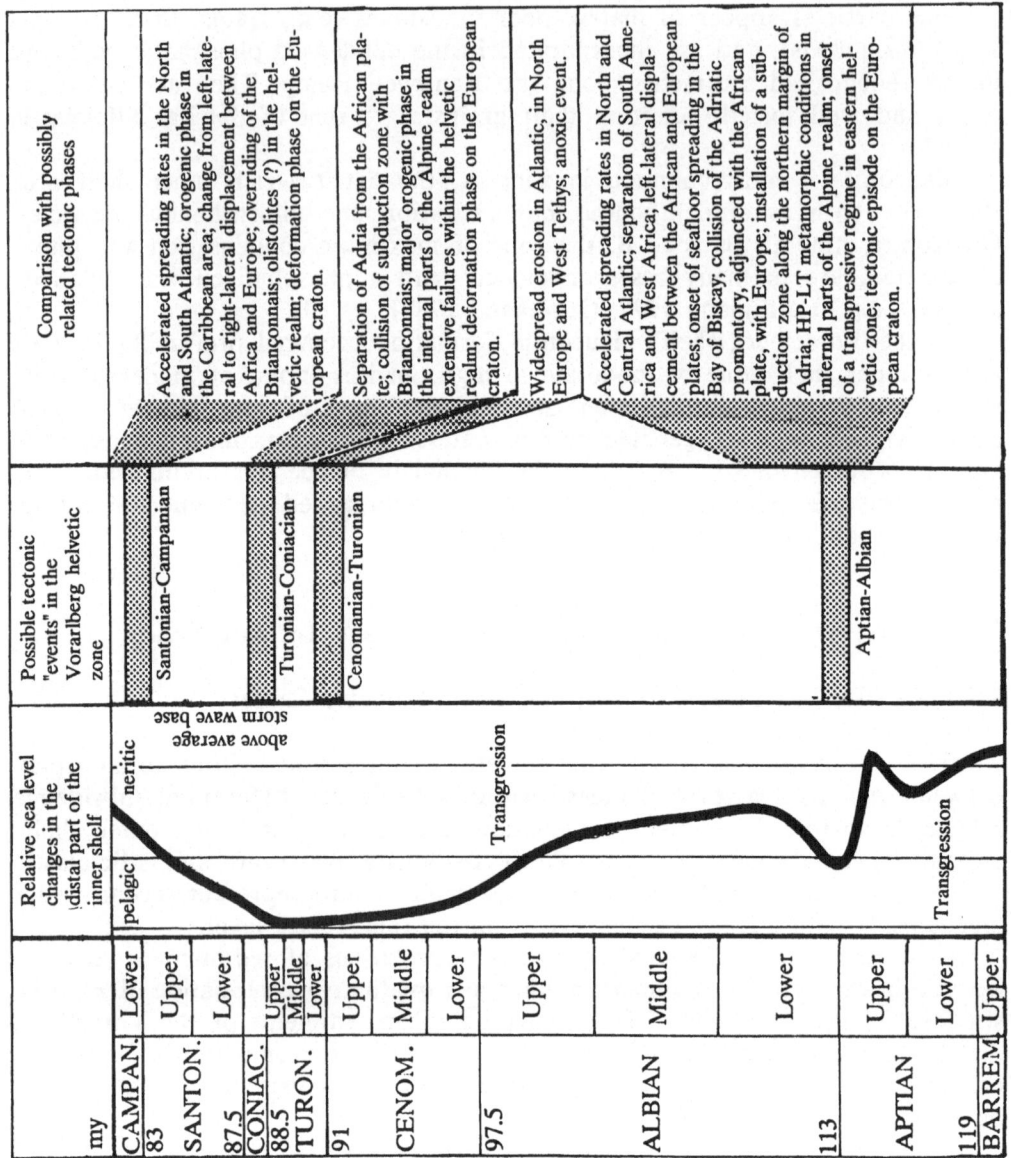

Fig. 55. Relative sea level changes in the Vorarlberg helvetic zone. Time distribution of possible tectonic events in the Vorarlberg helvetic zone and comparison with coeval regional and global tectonic phases (after Glangeaud and D'Albissin 1958; Juignet et al. 1973; Casey and Rawson 1973; Oberhauser 1973; Trümpy 1973, 1982, 1985; Vandenberg 1979; Schwan 1980; De Graciansky et al. 1981, 1986, 1987; Laubscher and Bernoulli 1982; Pindell and Dewey 1982; Smith and Woodcock 1982; Dercourt et al. 1985, 1986; Savostin et al. 1986; Ricou and Siddans 1986; Baird and Dewey 1986; Choukroune et al. 1986; Winkler and Lüdin 1986; Ziegler 1987, 1988; Flügel et al. 1987; Frank 1987; Le Pichon et al. 1988)

rather than a cause of the relative sea level fall, thus acting merely as an amplification factor. The seaward shift of the current could have been caused by a general eustatic sea level fall, as noted by Harris et al. (1984) and Haq et al. (1987). It could as well be caused by physiographic changes along the eastern part of the northern Tethys margin, deflecting the current in a seaward direction. The coeval opening of the Anglo-Parisian Basin toward the Boreal sea represents another influential factor, which had probably a twofold effect. The new influx of boreal water into the tethyan realm could have deflected the tethyan current system as well (Sect. 3.3; Fig. 44), and may have contributed to an observed regional deterioration in climate (during late Aptian and earliest Albian times; Juignet et al. 1973; Kemper 1983, 1987; Föllmi 1986; Bollinger 1986; Ouwehand 1987; cf. Vakhrameev 1978; Reyment and Bengston 1985, 1986), enhancing thus the availability and influx of siliciclastic detritus.

A relative sea level change in the latest Aptian represents probably the most dramatic change in the Mid- and early late Cretaceous sea level history of the helvetic shelf. In probably less than 1 my, a profound deepening occurred (Twäriberg Bed overlying Brisi Limestone). The deepening-upward trend was accompanied by pronounced differential subsidence between the inner and outer shelf, giving rise to the Rankweil Ramp with its characteristic broad and stable channels, and by widespread accumulation of redeposited sediments (Klaus and Rankweil Beds). The westbound tethyan current system was located in the proximal inner shelf, eroding larger areas and inducing phosphatization of sediments (Twäriberg Bed; Sect. 2.5).

A sea level rise of this magnitude is not indicated on the Haq et al. chart (1987; cf. also Harris et al. 1984). Relative sea level curves from other areas generally display regressive tendencies at the Aptian/Albian boundary (e.g., Jeletzky 1977; Matsumoto 1980; Naidin et al. 1980). Rapidity and amplitude of this relative sea level change, and the coeval phase of pronounced steepening of the Rankweil Ramp point to a tectonic "event" as the major cause of the relative sea level rise, probably related to an important "eo-alpine" orogenic phase (Fig. 55; Sect. 3.7.3).

Relative sea level changes during Albian and early Cenomanian times are indicated by lateral shifts in the position of zone B, the zone of prominent phosphogenesis, as well as by sequential landward shifts in the southward extension of siliciclastic sands (Sect. 3.6.2). A relative sea level fall in early Albian time was followed by a slow and persistent sea level rise, which continued into late Turonian time (Fig. 55). As a whole, the helvetic shelf has largely been stable during this period. Abrupt sea level changes are not detected, and crustal subsidence and eustatic sea level changes may have been balanced to a certain degree. This persistent, relative sea level rise corresponds to the worldwide observed, Mid- and early late Cretaceous eustatic sea level rise ("Cenoman-Transgression" of E. Suess 1883; e.g., Hancock and Kauffman 1979; Hallam 1984; Reyment and Bengston 1985, 1986).

In late Turonian time, conditions changed. The increasing benthic/planktonic ratios in preserved foraminifera and the stronger presence of calcispheres in upper sequences of the Seewen Formation indicate shallowing conditions, beginning in late Turonian or Coniacian time and persisting throughout the remaining Cretaceous (Figs. 3 and 55).

Turonian changes from deepening to shallowing trends or perturbances in continuing sea level rises are indicated in most sea level charts, although with inconsistent timing. A following persistent shallowing, however, is not known from other areas (Hancock and Kauffman 1979; Matsumoto 1980; Weimer 1984; Harris et al. 1984; Seiglie and Baker 1984; Haq et al. 1987).

Intriguingly, the late Turonian turning point in helvetic sea level history coincides with an episode of faulting and widespread failure (Upper Turonian-Coniacian Götzis Beds; Sect. 2.11). This event is probably related to a major eo-alpine orogenic phase that affected internal parts of the future Alpine realm (Fig. 55; Sect. 3.7.3).

Helvetic sea levels continued to fall throughout the remainder of the late Cretaceous. Large inner shelf areas emerged at the end of the Cretaceous (e.g., Trümpy 1982). Compressive stress, transmitted from the approaching eo-alpine orogenic front is probably the main cause of this persisting sea level drop (Sect. 3.7.3; Ziegler 1987).

3.7.2 Unconformities, Diachronism, and Condensation

Two types of unconformities are recognized within the Vorarlberg Mid- and upper Cretaceous sediment column: current-induced, multi-event unconformities, capable of transforming laterally into condensed beds, and single-event unconformities, related to the formation of gravity flow deposits. Transitions between these two types are common.

The first type is initiated by <u>current activity</u> and represents a genuine unconformity. Current-related unconformities are the result of a phase of erosion and/or nondeposition, and may transform laterally into condensed zones, which document very low sediment accumulation rates, caused by a dynamic interplay between deposition and erosion. Their extension is limited by topographic features (e.g., Rankweil Ramp), as well as by physical properties (current activity).

Current-related unconformities are the result of different events: (1) an initiating event, (2) zero to multiple unconformity maintaining events, and (3) an unconformity "closing" event. Evaluation of the character of events and the time span in between is useful for a better understanding of the unconformity (though not always possible):

1. Reconstruction of the <u>initiating event</u>: mapping of the unconformity basal plane by mapping the immediately underlying sediments and by evaluation of differences and tendencies in their age and thicknesses help to identify areas with stronger erosion (e.g., along the Rankweil Ramp, channels) and weaker erosion, and gives a semiquantitative estimation of eroded material (e.g., Figs. 32 and 34). Is the initiating event syn- or diachronous? Tectonic events are more likely to be synchronous (e.g., earthquakes), whereas shifting of a longshore current onto the shelf may represent a diachronous event (especially when coupled to a slow, relative sea level rise; Fig. 57).

Diachronous events give rise to time-transgressive unconformity bases (cf. Vail et al. 1977, 1984; Hallam 1984; Burton et al. 1987; Cross and Lessenger 1988). The presence of angular unconformities may point to a tectonic event as initiator, especially when the included time span is short (<1 my to several my).

2. Dating or estimation of the <u>time span included in the unconformity</u>, either by dating of under- and overlying sediments or by dating of fossils within a possibly included condensed bed. Lateral or proximal-distal differences in age are important. Two unconformities can unify, for instance (e.g., Fig. 2).

3. Reconstruction of the <u>terminating event</u>: is this event syn- or diachronous? Is it due to a relative sea level rise or fall (for example, the influx of siliciclastic sediments during a relative sea level fall)? Terminating events may also be accompanied by erosional events, which may cut through the basal unconformity plane (e.g., storm events during deposition of siliciclastic sands).

This class of unconformities is often related to sea level changes, but not *per se*: unconformities, invoked by current shifts, are not necessarily the result of sea level changes. Current shifts may also be the consequence of topographic changes along the current path.

A second type of unconformity is due to <u>single-event gravity flows</u>. This type does not truly represent an unconformity; it represents a truncation surface, which may mimic an unconformity, especially when it includes an artificial hiatus, induced by erosion of underlying sediments (true in most cases; cf. Shanmugam 1988).

Events, leading to the formation of erosive debris or high-density turbidity currents, induce widespread, more or less planar unconformities, which may include an artificial hiatus (depending on topography and physical properties of underlying sediment). The unconformity along the base of the *archaeocretacea* Zone Götzis Bed represents a good example of this type (Fig. 32 and 33). Large-scale shelf failures lead to irregular unconformities, including artificial hiatuses of varying time spans (e.g., at the base of the Upper Turonian-Coniacian Götzis Beds; Fig. 34). Tectonic events may lead to angular unconformities, overlain by widespread mass flows (this is probably true for the base of the Amden Formation).

This type of unconformity is the product of almost instantaneous processes, in most cases unrelated to sea level changes.

Current-induced, genuine unconformities may laterally evolve into single-event unconformities, related to gravity flow deposits. On the Vorarlberg helvetic shelf, this transformation takes place at the boundary of zone B, the zone of current-induced low sediment accumulation rates and major phosphogenesis, and zone C, the zone of redeposition, along the upper boundary of the Rankweil Ramp.

An important unconformity marks the boundary between Schrattenkalk and Drusberg Formations, and Garschella and Mittagspitz Formations (Fig. 2). On the Schrattenkalk platform, the unconformity embraces a hiatus, due to early to early late Aptian current-induced erosion in distal areas; to latest Aptian current-induced erosion in addition to this in intermediate areas; and to early and middle Albian current-induced erosion in addition to these in proximal areas. In the transition zone toward the outer shelf, the unconformity transforms into a condensed phosphatic bed (Luitere Bed). On the outer shelf, the Luitere Bed evolves into the Mittagspitz Formation, which displays a probable single-event unconformity at the base, overlain by a distinct, slump-folded bed (Felber and Wyssling 1979).

The ages of Schrattenkalk orbitolinids (Bollinger 1986) and phosphatized Luitere ammonoids indicate that the base of the inner shelf unconformity becomes younger toward the west and north. This is caused by the gradual landward shifting of the westbound current-system over the helvetic inner shelf (Figs. 56 and 57), inducing a gradual demise of the Schrattenkalk carbonate platform (Sects. 2.3.2 and 3.8). The genesis of this time-transgressive unconformity is closely related to a relative sea level rise. In distal inner shelf areas, the complex unconformity was closed by the influx of middle Upper Aptian siliciclastic Brisi and Gams Beds (commonly related with storm-induced erosive phases, enlarging the hiatus). In more proximal areas, the unconformity was either closed by younger siliciclastic beds, such as the Lower Albian Niederi Beds, the Lower to Middle Albian Sellamatt Beds or the Upper Albian Aubrig Beds, or it transformed proximally into Albian condensed beds, such as the Plattenwald Bed, including a large hiatus at the base (e.g., section 2 in Fig. 19).

A second important unconformity occurs between the middle Upper Aptian depositional sequence of Brisi Limestone, Brisi, Gams, and Freschen Beds, and the uppermost Aptian to lowermost Albian depositional sequence of Twäriberg, Klaus, and Rankweil Beds (Fig. 2). On the inner shelf, the unconformity consists of a current-induced erosional hiatus. It evolves into a condensed phosphatic bed in proximal directions (Twäriberg Bed), as well as in the interchannel areas along the Rankweil rampzone. The unconformity changes into a single-event unconformity at the base of the distal inner shelf Klaus Beds and the channel and fan systems along the Rankweil Ramp, where it includes an erosional hiatus. It disappears within the outer shelf Freschen and Hochkugel Beds. The unconformity has a conceivably synchronous base (although cut in sediments of different ages), and is probably the product of a tectonic event. Its formation is accompanied by a relative sea level rise (landward current shift onto the proximal inner shelf) and its end is marked by a relative sea level fall (Fig. 49).

The distal inner shelf condensed phosphatic Plattenwald bed of the Albian transforms into several unconformity-related condensed beds toward the north (e.g., on top of the Schrattenkalk Formation, the Niederi and Sellamatt Beds), which are induced by winnowing and erosion along the paired axes of a bifurcated westbound current system (Figs. 19 and 49). Along the Rankweil Ramp, the Plattenwald Bed transits into the fan and channel systems of the Rankweil Beds.

A fourth major unconformity separates the Garschella and Seewen Formations. It is characterized by an erosional hiatus and the presence of a condensed phosphatic bed in proximal parts of the inner shelf (Kamm Bed; Figs. 29

and 30), by a small hiatus in distal parts of the inner shelf, a major hiatus along the Rankweil Ramp (Fig. 2), and again a small but persistent hiatus in the outer shelf. The base of the unconformity is approximately synchronous (top of Freschen Beds, Plattenwald Bed, and Aubrig Beds); the top, however, is diachronous (due to diachronism within the basal Seewen Formation sediments; Fig. 2).

Unconformities appear also within the Seewen Formation, at the base of the *archaeocretacea* Zone Götzis Bed, respectively of the Upper Turonian-Coniacian Götzis Beds. They are due to major gravity flows, accompanied by erosion (Figs. 32 and 34), and are possibly triggered by tectonic events (Fig. 55; Sect. 3.7.3).

A major, commonly angular unconformity is present at the base of the Amden Formation. This unconformity is related to widespread erosion and the formation of extensive gravity flows, forming the base of the Amden Formation (Fig. 42). Again, the main cause may have been a tectonic event, corresponding to an orogenic phase in the internal Alpine realm (Fig. 55; Sect. 3.7.3).

3.7.3 The Possible Influence of "Eo-alpine" Tectonic Phases

Certain episodes in the Mid- and early late Cretaceous helvetic history are characterized by very rapid processes with a large impact on the configuration of the sediments. They are possibly related to phases of enhanced tectonic activity within the internal parts of the future Alpine realm (Fig. 55).

The latest Aptian-earliest Albian episode embodies a discrete episode of strong differential subsidence, rapid relative sea level rise, a major shift in the current system onto proximal parts of the inner shelf, the pronounced erosion and accumulation of larger quantities of redeposited sediments, and minor faulting in the inner shelf area. This event is probably correlatable to an important episode in the tethyan realm, in which the Adriatic promontory and the European plate collided (Fig. 55; 115-110 my event of many authors, although timing is uncertain).

Given the correctness of this correlation, this event indicates the onset of a transpressive regime in the helvetic shelf area, which induced a "basinal" phase (e.g., Ballance and Reading 1980; Hsü 1982; Trümpy 1982; Laubscher and Bernoulli 1982; Dercourt et al. 1985, 1986; Ziegler 1987; Frank 1987; Greber and Ouwehand 1988).

The *archeocretacea* Zone Götzis Bed, generated and deposited near the Cenomanian-Turonian boundary, documents a widespread phase of erosion and redeposition. This episode is difficult to judge. It is not correlatable with an eo-alpine tectonic phase (Fig. 55); however, it is coeval with the well-known Cenomanian-Turonian boundary anoxic event, which was preceded by widespread erosion (e.g., De Graciansky et al. 1981, 1986, 1987).

Arthur et al. (1988) considered this event as a major productivity event, perhaps triggered by a latest Cenomanian tectono-eustatic pulse, influencing deep-water circulation patterns.

The late Turonian-Coniacian episode is characterized by a large-scale failure event along the Rankweil Ramp, by faulting, and by the commencement of a relative sea level fall that persisted throughout the remainder of the Cretaceous. This event is traceable along the entire helvetic zone and marks the end of the "basinal" phase. It probably correlates to a widespread orogenic phase in the internal eo-alpine zone (Fig. 55; e.g., Glangeaud and D'Albissin 1958; Oberhauser 1973; Trümpy 1973, 1982, 1985; Schwan 1980; Föllmi 1981, 1986; Ziegler 1987; Frank 1987).

The episode of erosion, deposition of widespread gravity flows along the base of the Amden Formation, and faulting at the Santonian-Campanian boundary is correlated with an important phase in which left-lateral displacement between the Apulian and European plates switched into a right-lateral one (Fig. 55; Biju-Duval et al. 1977; Laubscher and Bernoulli 1982; Savostin et al. 1986; Flügel et al. 1987).

The duration of the "basinal" phase of the Vorarlberg helvetic area estimates approximately 25 my (latest Aptian to latest Turonian time). The "turning point" (= "inversion"; after Ziegler 1987) in latest Turonian is probably due to the gradual change from a transpressional into a more compressional regime along the eo-alpine front.

A remaining difficulty is represented by the unknown character of the transmission mechanism of stress from the active eo-alpine collision front onto the passive European borderland and through thinned continental crust of the Briançonnais and the Valais Trough, without being absorbed (e.g., Trümpy 1982, 1985). It is uncertain what kind of lever allowed the stress to span this distance through different rock types. A probable explanation is given with the assumption that the source of stress was not limited to the eo-alpine collision front, but that a smaller part of the stress was also delivered more directly via subplate boundaries, one of which is assumed to have been located within the Valais trough (Laubscher and Bernoulli 1982; cf. also Cloetingh 1986, 1988).

3.8 Drowning of the Carbonate Platform

Drowning of the Schrattenkalk shallow-water carbonate platform occurred in two steps, both closely related to changes in the flow pattern of the westbound current system (Figs. 56 and 57).

The first step was gradual and corresponded to the slow relative sea level rise in early and early late Aptian times. During this rise the current shifted onto the platform in a traceable pattern.

1. The carbonate production in the Allgäu helvetic zone (southeastern F.R.G.) ceased in late Barremian time. The Schrattenkalk Formation is covered with a condensed phosphatic bed enclosing ammonoids from the entire early Aptian to early late Aptian period (Luitere Bed; Gebhard 1983, 1985).
2. The carbonate production in the southeastern Vorarlberg helvetic shelf stopped prior to middle early Aptian (the Luitere Bed includes middle early

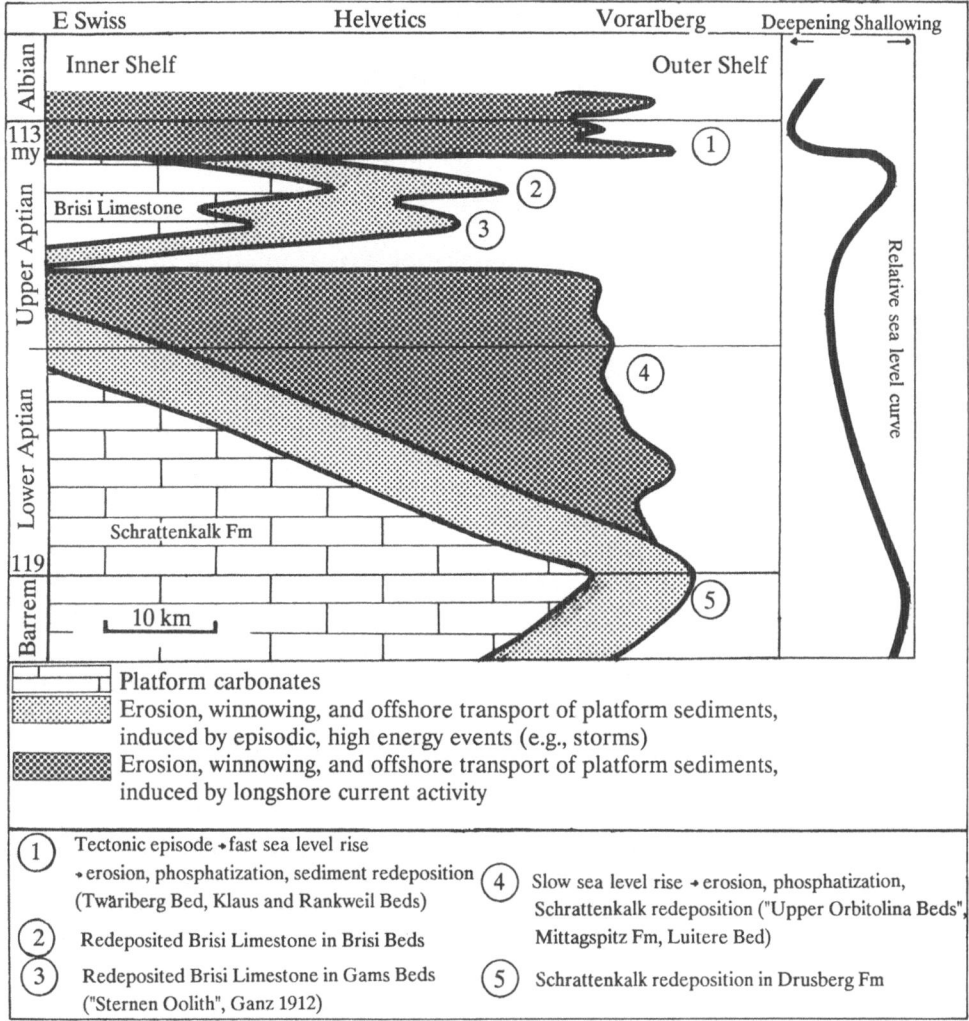

Fig. 56. Schematic overview of trends in drowning of the Schrattenkalk carbonate platform. Caution, this is not a time-stratigraphic diagram depicting the lithologic endproducts, but an overview of different processes that took place during this time interval.

Proximal shelf parts are represented in the central and eastern Swiss, distal shelf parts in the Vorarlberg and Allgäu helvetic areas. During the relative sea level rise in early Aptian time (slow, inducing a time-transgressive boundary) and at the Aptian-Albian boundary (fast, possibly triggered by a tectonic episode, inducing a "synchronous" boundary), a westbound, longshore, current system shifted onto the helvetic inner shelf and caused winnowing, erosion, and redeposition of sediments

to early late Aptian ammonoids). In western and northwestern areas of the Vorarlberg shelf, the carbonate production probably persisted throughout a major part of the early Aptian (based on orbitolinids; Bollinger 1986).

3. On the eastern and central Swiss portion of the carbonate platform, the Luitere ammonoids point to early late Aptian time. Orbitolinid dates suggest a continuing carbonate production throughout large parts of the early Aptian (Lienert 1965; Bollinger 1986).

In middle late Aptian time, the presence of Brisi Limestone records an ultimate progradational phase of the carbonate platform. This episode corresponds to a shift of the current system into outer shelf areas, which allowed the platform to regenerate itself (Fig. 57).

In latest Aptian time, a sudden, relative sea level rise, probably related to a tectonic event, provoked a major shift of the current system onto proximal areas of the shelf, thus terminating the carbonate production almost instantaneously (at least in the preserved, exposed part of the helvetic shelf).

Obviously, the shallow-water carbonate production ceased in locations, which were swept by the bottom-hugging tethyan current. The current water masses, probably enriched in suspended organic matter and somewhat reduced in oxygen content (Sects. 3.5 and 3.6), may have introduced an environment hostile to the specialized, oligotrophic, carbonate-producing organisms (algae, calcareous benthic foraminifera, calcareous sponges, stromatoporoids, hermatypic scleractinian corals, rudistic bivalves; e.g., Scholz 1979, 1984). This hostility may have been evoked by a combination of (1) erosive current strength; (2) episodic and insufficient availability of oxygen; (3) decreasing water transparency due to elevated organic matter contents; (4) excess in nutrients; (5) proliferation of rapidly growing microbial mats and soft-bodied algae; (6) aggravation of $CaCO_3$ precipitation in organic-rich environments; and (7) decoupling of zooxanthellid algae and corals (cf. Arthur and Schlanger 1979; Hallock and Schlager 1986; Hallock 1987).

A local (?) deterioration of climates in the late Aptian, and the new influx of boreal waters via the Anglo-Parisian Basin probably represented additional factors that contributed to the demise of the platform. An indicator for climatic deterioration is the different composition of calcareous fossils in the Brisi Limestone and the Schrattenkalk Formation. Hermatypic organisms (e.g., stromatoporoids and corals) are not preserved from the Brisi Limestone, and this is probably related to a general drop in temperature. Sedimentary structures in the Brisi Limestone (and in contemporaneous Brisi Beds) and the Brisi Limestone facies indicate high energy levels during their deposition, possibly related to the presence of frequent storms (Sects. 2.4.1 and 3.7.1). The change

Fig. 57. Shifts of the westbound tethyan current system onto and away from the carbonate platform (palinspastic map); indicated is the position of the current axis during earliest, middle early to late early, early late Aptian, middle late, and latest Aptian times

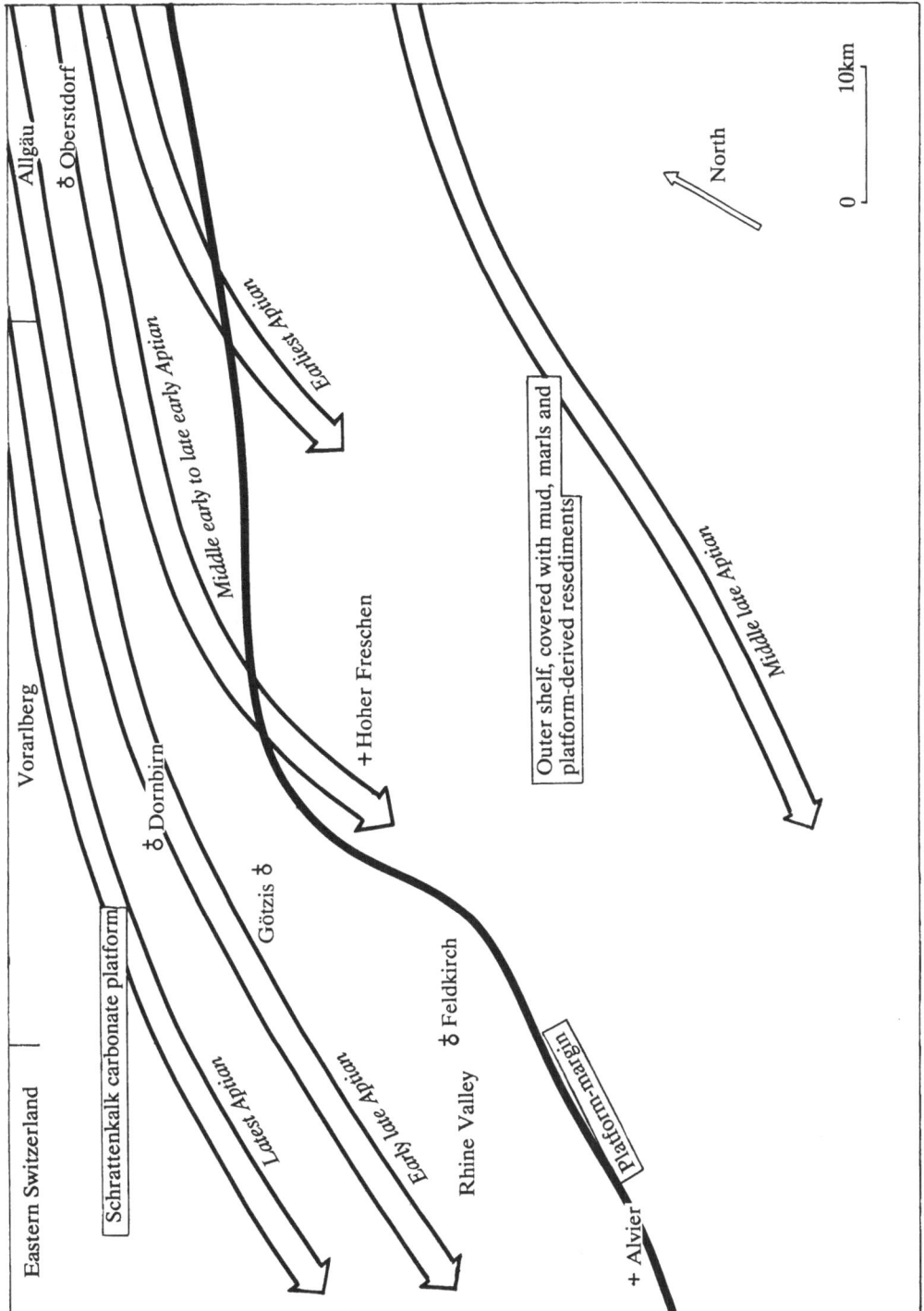

to a cooler, more humid climate during late Aptian time is noted by other investigators [helvetic area: Bollinger (1986) noted an increase in the illite/kaolinite ratio during late Aptian time; Ouwehand 1987; Tethys: Alvarez-Ramis et al. 1981; northern Europe: Juignet et al. 1973, Kemper 1983, 1987; cf. also Barron and Washington 1982; Barron et al. 1985].

Intriguingly, the demise of the Schrattenkalk carbonate platform occurred in a period of global platform drowning (Sect. 1.2; Schlager 1981). In the Aptian, intense global volcanic and tectonic activity induced high levels of CO_2 outgassing, and contributed probably to elevated atmospheric CO_2 levels (Berner et al. 1983; Berner and Lasaga 1989; Arthur et al. 1985). Simultaneously, the calcite compensation depth (CCD) within the oceans rose rapidly to a shallow level (Thierstein 1979; Tucholke and Vogt 1979). This certainly influenced the global $CaCO_3$ cycle, but it remains unclear to what extend the $CaCO_3$ production on shelves was affected [e.g., Arthur et al. (1985) suggested an inhibiting effect on the production of aragonite].

The demise of the Schrattenkalk carbonate platform seems therefore to be induced by the interaction of (1) shifting of the tethyan current system onto the carbonate platform, responsible for major erosion and flooding with waters, enriched in organic matter and somewhat reduced in oxygen content; (2) climate deterioration; (3) relative sea level changes; and (4) globally elevated CO_2-outgassing rates and a rapid rise in the CCD (cf. Schlager 1981; Kendall and Schlager 1981; Arthur et al. 1985; Hallock and Schlager 1986; Hallock 1987).

3.9 The Garschella Case and its Implications for Lower Cretaceous Cyclic Sedimentation

3.9.1 Early Cretaceous Cyclicity

Lower Cretaceous sediments of the helvetic shelf embody a suite of well-studied shallowing-upward sequences, consisting generally of a lower terrigenous and an upper calcareous part. Sequence thicknesses measure 200-400 m and time spans of deposition 2-5 my (Fichter 1934; Carozzi 1951; Brückner 1951, 1953; Lienert 1965; Funk 1971, 1985; Haldimann 1977; Strasser 1979, 1982; Trümpy 1980; Burger and Strasser 1981; Keller 1983; Burger 1985; Wyssling 1986; Bollinger 1986; Funk and Wyssling 1987; Fig. 3).

Sequence boundaries are marked by thin, condensed, and commonly phosphatized beds. These beds represent the "dark sides of the cycle". They document deepening-upward half-cycles, but their far from complete record, including major hiatuses, complicates reconstructions of the deepening-upward processes. The development of the condensed beds consumed more time on the average than the building up of the shallowing-upward sequences (2-7 my; Fig. 3).

Lower Cretaceous helvetic sediments resulted from cycles of comparatively rapid formation of thick, prograding, and shallowing-upward sequences, followed by longer periods of retarded sediment accumulation and phosphogenesis. Periodicities of the cycling process (4-8 my) are well beyond the Milankovitch range (20-400 ky; e.g., Barron et al. 1985; Goldhammer et al. 1987), but shorter than the average "periodicities" of Cambrian-Ordovician "Grand Cycles" of the North American craton (average duration of upward shallowing is 9-15 my; e.g., Chow and James 1987). They may be comparable to the timescale of late Paleozoic and Mesozoic sedimentary and climatic cycles (e.g., Ross and Ross 1985; Bayer and McGhee 1987; Schlanger 1986; Rampino and Stothers 1987).

Most helvetic investigators proposed an "autocyclic" mechanism in terms of episodically reinforced subsidence (cf. Fichter 1934). Others pointed to the importance of allocyclic mechanisms such as sea level and climatic changes (e.g., Brückner 1951, 1953; Strasser 1982). Based on geohistory diagrams, Funk (1985) invoked a mechanism dominated by subsidence pattern (compare also Funk and Wyssling 1987).

3.9.2 Mid-Cretaceous Distortion

At the end of the early Cretaceous, the rather regular process of cyclicity was distorted. Barremian to Lower Aptian Schrattenkalk shallow-water carbonates are covered by a comparatively thin sequence consisting of phosphates (Luitere Bed), sandstones (Brisi Beds), as well as shallow-water carbonates (Brisi Limestone). This sequence represents an ultimate, somewhat rudimentary, shallowing-upward sequence within the series of Lower Cretaceous cycles, which was obviously aborted by the Aptian-Albian boundary episode encompassing a rapid relative sea level rise and a strong landward current shift. The overlying, strongly condensed Albian phosphatic beds do not lead into a new shallowing-upward sequence. Instead, they heralded a pelagic regime that ruled the helvetic shelf from latest Albian onwards (Fig. 3).

The interruption of the early Cretaceous cyclicity process is probably caused by the installation of a transpressive regime, which induced the "basinal phase" in the helvetic zone from the latest Aptian onwards. The transpressional and subsequent compressional forces overprinted the subtle mechanism of cyclicity in such a way that it was overwhelmed.

3.9.3 The Tethyan Current System as an Influential Factor on Cyclicity

A comparison of Lower Cretaceous condensed phosphatic beds of the helvetic shelf with those of the Mid-Cretaceous Garschella Formation reveals similarities, such as the mode of accumulation (presence of multi-event winnowed, condensed, phosphatic beds), degree of condensation (sediment accumulation rates <10 cm/my), and occurrence in zones approximately parallel to the shelf edge. The similarity of these configurations suggests that the tethyan current system was actively involved in the formation of the lower Cretaceous phosphatic beds as well. Various authors pointed, indeed, to a relationship between the development of Lower Cretaceous condensed beds and currents (Heim in

Baumberger et al. 1907; Heim and Baumberger 1933; Heim 1958; Funk 1971; Haldimann 1977; Wyssling 1986).

The westbound tethyan current may have played an active role in the demise of the former carbonate production sites, represented by the Oehrli, Betlis, and Kieselkalk Formations (Fig. 3), by means of episodic flooding and eroding of proximal helvetic shelf areas with probably fertile and partly oxygen-depleted bottom waters. Landward current shifts, terminating platform-carbonate production and inducing low sediment accumulation rates and phosphogenesis, and seaward current shifts, allowing for the influx of terrigenous sediments, progradation, and the final build-up of carbonate platforms, alternated in a rhythmic, cyclic pattern, which probably resulted from apparently rhythmic relative sea level changes and regular regional tectonic subsidence patterns.

4 Conclusions and Summary

This study adds pieces to the reconstruction of the puzzling paleoceanographic evolution of the Mid- and lower Upper Cretaceous helvetic shelf, situated along the southern European, northern Tethys margin. It shows the development of a Barremian-Lower Aptian shallow-water carbonate platform, its drowning, the development of condensed phosphatic beds on the current-dominated, drowned platform, the subsequent installation of a pelagic regime and the final return to shallower waters. This work suggests the following events and episodes as important stations along the evolutionary path of the Mid- and early late Cretaceous helvetic shelf:

1. **Barremian and early Aptian.** Development of an extensive shallow-water carbonate platform; oligotrophic waters; diversification of specialized carbonate-producing organisms; development of oolitic shoals; periodic, offshore transport of carbonate particles over a gentle, homoclinal ramp into the outer muddy shelf, possibly by storm events.

2. **Early and early late Aptian.** First step in the drowning process of the carbonate platform; slow relative sea level rise; diachronous development of condensed, phosphatic beds, signalizing the slow shift of the westbound tethyan current system onto the platform; installation of an oxygen minimum zone, limited in its upward extension by the current system; current erosion of the platform and flooding with water masses, probably enriched in organic matter and impoverished in oxygen, which inhibited the carbonate production; current-induced phosphatization and condensation by means of catastrophic burial of biotic communities with mobile palimpsest sands and subsequent winnowing; offshore transport of winnowed and eroded sediments onto the outer shelf *via* channels until equilibrium was reached, i.e., all palimpsest sands were removed; differential subsidence along the platform margin, due to different compaction rates of inner, dominantly calcareous and outer, dominantly muddy and hemipelagic sediments; dominance of ammonoids with tethyan character and with close affinities to eastern European faunas.

3. **Middle late Aptian.** Relative sea level fall; climate deterioration; opening of a seaway *via* the Anglo-Parisian Basin to the Boreal realm; influx of boreal waters into Tethys; seaward shift of the tethyan current system towards the outer shelf, where it acted as a contour current, resuspending and redistributing fine-grained sediments; ultimate progradation of shallow-water carbonates, produced by an impoverished calcareous fauna (dominated by crinoids; absence of hermatypic organisms); possibly storm-dominated distribution of coarse-grained siliciclastic sands on the inner shelf; episodic, offshore transport of sands along mobile channel and fan systems onto the outer shelf; deposition of laminated marls and clays on the outer shelf.

4. **Latest Aptian and earliest Albian.** Tectonic event and onset of a transpressive regime in the helvetic zone, inducing the "basinal" phase, and terminating early Cretaceous cyclicity; rapid relative sea level rise; reinforced dif-

ferential subsidence between the inner and outer shelf, resulting in the formation of a distally steepened ramp (Rankweil Ramp); major current shift onto proximal areas of the inner shelf, inducing erosion and phosphatization; final, rapid step in the drowning of the platform, due to flooding with probably fertile waters, to strongly erosive conditions, to deteriorated climatic conditions, and to a rapid, relative sea level rise; transport of eroded sediments into distal areas of the inner shelf and, *via* newly formed, broad, and stable channels along the Rankweil Ramp, into proximal areas of the outer shelf; phosphatization and condensation in interchannel areas on the Rankweil Ramp; continuing deposition of laminated clays and marls on the outer shelf. Dominance of boreal and eurythermic ammonoids.

5. **Early and late Albian.** Relative sea level fall (early Albian), followed by a slow but persistent relative sea level rise; current bifurcation, leading to two parallel, coeval zones of phosphogenesis and very low accumulation rates; three phases in which siliciclastic sands were replenished (middle early, late early, and middle late Albian); zones along current axes needed ca. 1.5 my to come into equilibrium, i.e., until all palimpsest sands were removed; episodic dislocation of the palimpsest sands caused catastrophic burial of entire biotic communities, enabling phosphatization within these sands; transport of spillover sediments *via* channels along the Rankweil Ramp into the proximal outer shelf; continuing deposition of laminated marls and clays on the outer shelf; tethyan and boreal ammonoids during early and middle Albian, cosmopolitan ammonoids during late Albian.

6. **Latest Albian and early Cenomanian.** Continuing relative sea level rise, persisting into the late Turonian; shift of current system and with it, the zone of low accumulation rates and phosphatization onto the proximal inner shelf; deposition of pelagic oozes on the rest of the shelf, with the exception of the Rankweil Ramp, which became a site of zero sediment accumulation rates and slight erosion (channels are infilled and deactivated).

7. **Middle and late Cenomanian.** Definitive installation of the pelagic regime in the entire exposed, preserved helvetic zone; continuing zero sediment accumulation rates along Rankweil Ramp.

8. **Latest Cenomanian and earliest Turonian.** An tectonic (?) event in the proximal inner shelf triggered the release of a debris flow or high-density turbidity current, causing the distribution of eroded glauconitic sand over large areas of the inner and outer shelf.

9. **Middle Turonian.** Continuing deposition of pelagic oozes.

10. **Late Turonian and early Coniacian.** Tectonic event; change from a relative sea level rise to a relative sea level fall (persistent throughout the remaining Cretaceous); end of the basinal phase ("inversion"); major failure along the Rankweil Ramp, resulting in the widespread occurrence of debris flows and megaturbidites including sediments as old as Barremian.

11. **Coniacian to late Santonian.** Continuing deposition of pelagic oozes.

12. **Late Santonian to early Campanian.** Tectonic event; widespread gravity flows of silts and muds, including larger slides of older sediments; deposition of pelagic oozes limited to distal outer shelf areas.

Dynamic events and episodes determined and directed the evolution of the Mid-Cretaceous triad of shallow-water carbonates, condensed phosphatic and glauconitic beds, and pelagic carbonates on the Vorarlberg Helvetic shelf, along the northern Tethys margin (Ager 1973, 1984). Directly or indirectly related regional dynamics, such as current-activity, relative sea level changes, tectonic events, and climatic changes, induced by the globally enhanced tectonic forces, as well as by orbital parameters, shaped this Mid-Cretaceous sequence, by defining the physical, and influencing the chemical and biologic conditions, under which sedimentary processes took place. Resulting dynamic processes on a short-term, stochastic base, such as slope failure, triggering of turbidity currents, and faulting, or on a long-term, internally quasi-periodic base, such as current-induced winnowing and condensation, differential subsidence, and drowning of carbonate platforms, were the prime and most influential "printers" of the preserved Mid-Cretaceous sedimentary document, and, in the same moment, the destroyers of many pages, leaving an obviously incomplete book behind. However, the more incomplete such sedimentary documents are, the more one's attention is drawn to the dynamics, responsible for such hiati. This study, therefore, is an attempt to read between the lines and to demystify some of the dynamics, responsible for the "missing pages".

References

Ager DV (1973) The nature of the stratigraphical record. Wiley, New York, pp 114

Ager DV (1984) The stratigraphic code and what it implies. In: Berggren WA, Van Couvering JA (eds) Catastrophes and earth history. Princeton Univ Press, pp 91-100

Alvarez-Ramis C, Biondi E, Desplato D, Hughes NF, Koenigner JC, Pons D, Rioult M (1981) Les végétaux (macrofossils) du Crétacé moyen de l'Europe occidentale et du Sahara. Végétations et paléoclimats. Cret Res, 2: 339-359

Arnaud-Vanneau A (1980) Micropaléontologie, paléoécologie et sedimentation d'une plate-forme carbonatée de la marge passive de la Téthys: L'Urgonien du Vercors septentrional et de la Chartreuse. Géol Alp Mém, 11/1-3, pp 874

Arthur MA, Dean WE, Pratt LM (1988) Global ocean-atmosphere geochemical and climatic effects of the Cenomanian/Turonian marine productivity event. EOS, (abstr) 69/16: 300

Arthur MA, Dean WE, Schlanger SO (1985) Variations in the global carbon cycle during the Cretaceous related to climate, volcanism, and changes in atmospheric CO_2. In: Sundquist ET, Broecker WS (eds) The carbon cycle and atmospheric CO_2: natural variations Archean to present. Am Geophys Union, Geophys Monogr, 32: 504-529

Arthur MA, Jenkyns HC (1981) Phosphorites and paleoceanography. Oceanol Acta Spec Nr, pp 83-96

Arthur MA, Schlanger SO (1979) Cretaceous "oceanic anoxic events" as causal factors in development of reef-reservoired giant oil fields. Am Assoc Pet Geol Bull, 63/6: 870-885

Arthur MA, Schlanger SO, Jenkyns HC (1987) The Cenomanian-Turonian oceanic anoxic event, II. Paleoceanographic controls on organic-matter production and preservation. In: Brooks J, Fleet AJ (eds) Marine petroleum source rocks. Geol Soc Lond Spec Publ, 26: 401-420

Baird AW, Dewey JF (1986) Structural evolution in thrust belts and relative plate motion: the upper Pennine Piemont zone of the internal Alps, southwest Switzerland and northwest Italy. Tectonics, 5/3: 375-387

Ballance PF, Reading HG (1980) Sedimentation in oblique-slip mobile zones. Int Assoc Sediment Spec Publ, 5, pp 265

Barron EJ, Washington WM (1982) Cretaceous climate: a comparison of atmospheric simulations with the geologic record. Palaeogeogr Palaeoclimatol Palaeoecol, 40: 103-133

Barron EJ, Arthur MA, Kauffman EG (1985) Cretaceous rhythmic bedding sequences: a plausible link between orbital variations and climate. Earth Planet Lett, 72: 327-240

Baturin GN (1971) Stages of phosphorite formation on the ocean floor. Nat Phys Sci, 232: 61-62

Baturin GN (1982) Phosphorites on the sea floor. Dev sediment, Elsevier, Amsterdam, 33, pp 343

Baumberger E, Heim A, Buxtorf A (1907) Paläontologisch-stratigraphische Untersuchung zweier Fossilhorizonte an der Valanginien-Hauterivian Grenze im Churfirsten-Mattstock Gebiet. Abh Schweiz Paläont Ges, 34/2: 1-30

Bayer U, McGhee GR (1987) Cyclic patterns in the Paleozoic and Mesozoic: implications for time scale calibrations. Paleoceanography, 1/4: 383-402

Bentz F (1948) Geologie des Sarnersee-Gebietes (Kt. Obwalden). Eclogae Geol Helv, 41/1: 1-77

Berggren WA (1982) Role of ocean gateways in climatic change. In: Berger WH, Crowell JC (eds) Climate in earth history. Nat Academic Press, Washington, pp 118-125

Berner RA (1982) Burial of organic carbon and pyrite sulfur in the modern ocean: its geochemical and environmental significance. Am J Sci, 282: 451-473

Berner RA, Lasaga AC (1989) Modeling the geochemical carbon cycle. Sci Am, 260/3: 74-81

Berner RA, Lasaga AC, Garrels RM (1983) The carbonate-silicate geochemical cycle and its effect on atmospheric carbon dioxide over the past 100 million years. Am J Sci, 283: 641-683

Biju-Dival B, Dercourt J, Le Pichon X (1977) From the Tethys ocean to the mediterranean seas; a plate tectonic model of the evolution of the western Alpine system. In: Biju-Duval B, Montadert L (eds) Structural history of the Mediterranean basins. Edit Technips, Paris, pp 143-164

Birch GF (1979) Phosphatic rocks on the western margin of South Africa. J Sediment Pet, 49: 93-110

Birch GF (1980) A model of penecontemporaneous phosphatization by diagenetic and authigenic mechanisms from the western margin of southern Africa. Soc Econ Paleontol Mineral Spec Publ, 29: 79-100

Blumer E (1905) Der östliche Teil des Säntisgebirges. Beitr Geol Karte Schweiz NF, 16/3: 518-638

Bolli HM (1944) Zur Stratigraphie der oberen Kreide in den höheren helvetischen Decken. Eclogae Geol Helv, 37/2: 217-328

Bollinger D (1986) Die Entwicklung des distalen osthelvetischen schelfes in Barremian und Unter-Aptian (Drusberg-, Schrattenkalk- und Mittagspitz-Formation). PhD Thesis, ETH Zürich, pp 136

Bouma AH (1987) Megaturbidite: an acceptable term? Geo-Marine Lett, 7: 63-67

Bourrouilh R (1987) Evolutionary mass flow - megaturbidites in an interplate basin: example of the North Pyrenean basin. Geo-Marine Lett, 7: 69-81

Bralower TJ (1988) Calcareous nannofossil biostratigraphy and assemblages of the Cenomanian-Turonian boundary interval: implications for the origin and timing of oceanic anoxia. Paleoceanography, 3/3: 275-316

Bralower TJ, Thierstein HR (1984) Low productivity and slow deep-water circulation in Mid-Cretaceous oceans. Geology, 12: 614-618

Bralower TJ, Thierstein HR (1987) Organic-carbon and metal accumulation in Holocene and Mid-Cretaceous marine sediments: palaeoceanographic significance. In: Brooks J, Fleet AJ (eds) Marine petroleum source rocks. Geol Soc Lond Spec Publ, 26: 345-371

Brandt K (1985) Sea level changes in the Upper Sinemurian and Pliensbachian of southern Germany. In: Bayer U, Seilacher A (eds) Sedimentary and evolutionary cycles. Lect Notes Earth Sci, Springer, Berlin Heidelberg New York Tokyo, 1: 113-126

Brass GW, Southam JR, Peterson WH (1982) Warm saline bottom water in the ancient ocean. Nature (London), 296: 620-623

Bréhéret JG, Caron M, Delamette M (1986) Niveaux rich en matière organique dans l'Albien vocontien; quelques caractères du paléoenvironment; essai d'interprétation genètique. Doc Bur Rech Géol Miner, 110: 141-191

Brett CE, Seilacher A (in press) Obrution deposits: a taphonomic consequence of event sedimentation. In: Einsele G, Ricken W, Seilacher A (eds) Cycles and events in stratigraphy. Springer, Berlin Heidelberg New York Tokyo

Broecker WS, Peng TH (1982) Tracers in the sea. Lamont-Doherty Geol Observ, Palisades NY, pp 690

Brückner WD (1951) Lithologische Studien und zyklische Sedimentation in der helvetischen Zone der Schweizeralpen. Geol Rundschau, 39/1: 196-216

Brückner WD (1953) Cyclic calcareous sedimentation as an index of climatic variations in the past. J Sediment Pet, 23: 235-237

Burger H (1985) Palfries-Formation, Öhrli-Formation und Vitznau-Mergel (basale Kreide des Helvetikums) zwischen Reuss und Rhein. PhD Thesis, Univ Zürich, pp 237

Burger H, Strasser A (1981) Lithostratigraphische Einheiten der untersten helvetischen Kreide in der Zentral- und Ostschweiz. Eclogae Geol Helv, 74/2: 529-560

Burnett WC (1977) Geochemistry and origin of phosphorite deposits from off Peru and Chile. Geol Soc Am Bull, 88: 813-823

Burnett WC (1980) Apatite-glauconite associations off Peru and Chile: palaeooceanographic implications. J Geol Soc Lond, 137: 757-764

Burnett WC, Beers MJ, Roe KK (1982) Growth rates of phosphate nodules from the continental margin off Peru. Science, 215: 1616-1618

Burnett WC, Roe KK (1983) Upwelling and phosphorite formation in the ocean. In: Suess E, Thiede J (eds) Coastal upwelling, its sediment record. Plenum Press, New York, Ser A, pp 377-397

Burnett WC, Veeh HH, Soutar A (1980) U-series, oceanographic and sedimentary evidence in support of recent formation of phosphate nodules off Peru. In: Bentor YK (ed) Marine phosphorites. Soc Econ Paleontol Mineral Spec Publ, 29: 61-72

Burton R, Kendall C, Lerche I (1987) Out of our depth; on the impossibility of fathoming eustacy from the stratigraphic record. Earth Sci Rev, 24: 237-277

Carbone S, Grasso M, Lentini F, Pedley HM (1987) The distribution and palaeoenvironment of early Miocene phosphorites of southern Sicily and their relationships with the Maltese phosphorites. Palaeogeogr Palaeoclimatol Palaeoecol, 58: 35-53

Caron M (1985) Cretaceous planktic foraminifera. In: Bolli HM, Saunders JB, Perch-Nielsen K (eds) Plankton stratigraphy. Cambridge Univ Press, pp 17-86

Carozzi A (1951) Rythmes de sédimentation dans le Crétacé helvétique. Geol Rundschau, 39/1: 177-195

Casey R (1961) The stratigraphical palaeontology of the Lower Greensand. Palaeontology, 3/4: 487-621

Casey R, Rawson P (1973) The Boreal Lower Cretaceous. Geol J Spec Issues, 5, pp 448

Cayeux L (1936) Existence de nombreuses bactéries dans les phosphates sediments de tout âge: consèquences. CR Acad Sci, 203: 1198-1200

Choukroune M, Ballèvre M, Cobbold P, Gautier Y, Merle O, Vuichard JP (1986) Deformation and motion in the western Alpine arc. Tectonics, 5/2: 215-226

Chow N, James NP (1987) Cambrian grand cycles: a northern Appalachian perspective. Geol Soc Am Bull, 98: 418-429

Cloetingh S (1986) Intraplate stresses, sea level changes and the sedimentary record. Int Assoc Sediment, Canberra, (abstr) pp 63-64

Cloetingh S (1988) Intraplate stresses: a new element in basin analysis. In: Kleinspehn KL, Paola C (eds) New perspectives in basin analysis. Springer, Berlin Heidelberg New York Tokyo, pp 205-230

Coleman JM (1988) Dynamic changes and processes in the Mississippi river delta. Geol Soc Am Bull, 100/7: 999-1015

Coleman JM, Prior DB (1980) Deltaic sand bodies. Am Assoc Pet Geol Cont Educ Course Note Ser, 15, pp 171

Cross TA, Lessenger MA (1988) Seismic stratigraphy. Annu Rev Earth Planet Sci, 16: 319-354

Crowley TJ, North GR (1988) Abrupt climate change and extinction events in earth history. Science, 240: 996-1002

Culvin SJ, Brunner CA, Nittrouer CA (1988) Observations of a fast burst of the deep western boundary undercurrent and sediment transport in south Wilmington Canyon from DSRV *Alvin*. Geo-Marine Lett, 8: 159-165

Dahanayake K, Krumbein WE (1985) Ultrastructure of a microbial mat-generated phosphorite. Mineral Deposita, 20: 260-265

De Graciansky PC, Bourbon M, Lemoine M, Sigal J (1981) The sedimentary record of Mid-Cretaceous events in the western Tethys and central Atlantic Oceans and their continental margins. Eclogae Geol Helv, 74/2: 353-367

De Graciansky PC, Deroo G, Herbin JP, Jaquin T, Magniez F, Montadert I, Müller C, Ponsot C, Schaaf A, Sigal J (1986) Ocean-wide stagnation episodes in the late Cretaceous. Geol Rundschau, 75/1: 17-41

De Graciansky PC, Brosse E, Deroo G, Herbin JP, Montadert I, Müller C, Sigal J, Schaaf A (1987): Organic-rich sediments and palaeoenvironmental reconstructions of the Cretaceous North Atlantic. In: Brooks J, Fleet AJ (eds) Marine petroleum source rocks. Geol Soc Lond Spec Publ, 26: 317-344

Delamette M (1981) Sur la découverte de stromatolithes circalittoraux dans la partie moyènne du Crétacé nordsubalpin (Alpes occidentales françaises). CR Acad Sci, 292: 761-764

Delamette M (1985) Phosphorites et paléocéanographie: l'Exemple des phosphorites du Crétacé moyen delphino-helvétique. CR Acad Sci, 300: 1025-1028

Delamette M (1988a) L'Evolution du domaine helvétique (entre Bauges et Morcles) de l'Aptien supérieur au Turonien: séries condensées, phosphorites et circulations océaniques. Publ Dep Géol Paléont Univ Geneva, 5, pp 316

Delamette M (1988b) Relation between the condensed Albian deposits of the helvetic domain and the oceanic current-influenced continental margin of the northern Tethys. Bull Soc Géol France, 8/IV/5: 739-745

Delamette M, Föllmi KB, Ouwehand PJ (1984) Occurrence of deep-water stromatolites at the boundary from Lower to Upper Cretaceous in Delphino-Helvetic units (F, CH, A). Schweiz Natf Ges, (abstr) 164: 42

Delaney ML, Boyle EA (1988) Tertiary paleoceanic chemical variability: unintended consequences of simple geochemical models Paleoceanography, 3/2: 137-156

Dercourt J, Zonenshain LP, Ricou LE, Kazmin VG, Le Pichon X, Knipper AL, Grandjacquet C, Sborshchicov IM, Boulin J, Sorokhtin O, Geyssant J, Lepvrier C, Biju-Duval B, Sibuet JC, Savostin LA, Westphal M, Lauer JP (1985) Présentation de 9 cartes paléogéographique au 1/20.000.000 s'étendant de l'Atlantique au Pamir pour la période du Lias à l'actuel. Soc Géol France, Bull, 8/I/5: 637-652

Dercourt J, Zonenshain LP, Ricou LE, Kazmin VG, Le Pichon X, Knipper AL, Grandjacquet C, Sbortshikov IM, Geyssant J, Lepvrier C, Pechersky DH, Boulin J, Sibuet JC, Savostin LA, Sorokhtin O, Westphal M, Bazhenov ML, Lauer JP, Biju-Duval B (1986) Geological evolution of the Tethys belt from the Atlantic to the Pamirs since the Lias. Tectonophysics, 123: 241-315

Dott RH Jr (1988) Something old, something new, something borrowed, something blue - a hindsight and foresight of sedimentary geology. J Sediment Pet, 58/2: 358-364

Druschtchitz VV, Gorbatschik TN (1979) Zonengliederung der Unteren Krei-de der südlichen UdSSR nach Ammoniten und Foraminiferen. In: Wiedmann J (ed) Aspekte der Kreide Europas. Int Union Geol Sci, Ser A, 6: 107-116

Dzulynski S, Ksiazkiewicz M, Kuenen PH (1959) Turbidites in flysch of the Polish Carpathian Mountains. Geol Soc Am Bull, 70: 1089-1118

Einsele G, Ricken W, Seilacher A (in press) Cycles and events in stratigraphy, general concepts and nomenclature. In: Einsele G, Ricken W, Seilacher A (eds) Cycles and events in stratigraphy. Springer, Berlin Heidelberg New York Tokyo

Felber P, Wyssling G (1979) Zur Stratigraphie und Tektonik des Südhelveti-kums im Bregenzerwald (Vorarlberg). Eclogae Geol Helv, 72/3: 673-714

Fichter H (1934) Geologie der Bauen-Brisen-Kette am Vierwaldstättersee und die zyklische Gliederung der Kreide und des Malm der helvetischen Decken. Beitr Geol Karte Schweiz NF, 69, pp 128

Filipek LH, Owen RM (1981) Diagenetic controls of phosphorus in outer continental-shelf sediments from the Gulf of Mexico. Chem Geol, 33: 181-204

Fischer A (1981) Climatic oscillations in the biosphere. In: Nitecki MH (ed) Biotic crises in ecological and evolutionary time. Academic Press, London New York, pp 103-131

Fischer A (1984) The two phanerozoic cycles. In: Berggren WA, Van Couvering JA (eds) Catastrophies and earth history. Princeton Univ Press, pp

129-150

Flemming BW (1981) Factors controlling shelf sediment dispersal along the southeast African continental margin. Mar Geol, 42: 259-277

Flemming BW (1988) Pseudo-tidal sedimentation in a non-tidal shelf environment (southeast African continental margin). In: De Boer PL, Van Gelder A, Nio SD (eds) Tide-influenced sedimentary environments and facies. Reidel, Dordrecht, pp 167-180

Flügel HW, Faupl P, Mauritsch HJ (1987) Implications on the Alpidic evolution of the eastern parts of the eastern Alps. In: Flügel HW, Faupl P (eds) Geodynamics of the eastern Alps. Deuticke, Vienna, pp 407-418

Föllmi KB (1981) Sedimentäre Hinweise auf oberkretazische Tektonik im Vorarlberger Helvetikum. Eclogae Geol Helv, 74/1: 175-187

Föllmi KB (1986) Die Garschella- und Seewer Kalk-Formation (Aptian-Santonian) im Vorarlberger Helvetikum und Ultrahelvetikum. PhD Thesis, ETH Zürich, Mitt Geol Inst ETH Univ Zürich NF, 262, pp 392

Föllmi KB (1988) Phosphate maximum zone. Am Assoc Pet Geol Bull, (abstr) 72/2: 186

Föllmi KB (1989 in press) Beschreibung neugefundener Ammonoidea aus der Vorarlberger Garschella-Formation (Aptian-Albian). Jahrb Geol Bundesanst Vienna, 132/1.

Föllmi KB, Garrison RE, Grimm KA (in press) Stratification in phosphatic sediments: illustrations from the Neogene of Central California. In: Einsele G, Ricken W, Seilacher A (eds) Cycles and events in stratigraphy. Springer, Berlin Heidelberg New York Tokyo

Föllmi KB, Garrison RE, Ramirez PC, Zambrano-Ortiz F (1988) Late Cretaceous phosphorite sequences in the central Colombian Andes. Int Geol Corr Progr, 156 Phosphorites, Oxford, (Abstr) p 9

Föllmi KB, Ouwehand PJ (1987) Garschella-Formation und Götzis-Schichten (Aptian-Coniacian): Neue stratigraphische Daten aus dem Helvetikum der Ostschweiz und des Vorarlbergs. Eclogae Geol Helv, 80/1: 141-191

Frakes LA, Francis JE (1988) A guide to Phanerozoic cold polar climates from high-latitude ice-rafting in the Cretaceous. Nature (London), 333: 547-549

Frank W (1987) Evolution of the Austroalpine elements in the Cretaceous. In: Flügel HW, Faupl P (eds) Geodynamics of the eastern Alps. Deuticke, Vienna, pp 379-406

Froelich PN (1984) Interactions of the marine phosphorus and carbon cycles. In: Moore B, Dastoor MN (eds) The interaction of global biochemical cycles. Jet Propulsion Lab, NASA Publ, 84-21: 141-176

Froelich PA, Arthur MA, Burnett WC, Deakin M, Hensley V, Jahnke R, Kaul L, Kim KH, Roe K, Soutar A, Vathakanon C (1988) Early diagenesis of organic matter in Peru continental margin sediments; phosphorite precipitation. Mar Geol, 80: 309-343

Froelich PN, Kim KH, Jahnke RA, Burnett WC, Soutar A, Deakin M (1983) Pore water fluoride in Peru continental margin sediments. Geochim Cosmochim Acta, 47: 1605-1612

Funk H (1971) Zur Stratigraphie und Lithologie des Helvetischen Kieselkalkes und der Altmannschichten in der Säntis-Churfirsten-Gruppe (Nordostschweiz).

Eclogae Geol Helv, 64/2: 345-433

Funk H (1985) Mesozoische Subsidenzgeschichte im helvetischen Schelf der Ostschweiz. Eclogae Geol Helv, 78/2: 249-272

Funk H, Briegel U (1979) Le facies Urgonien des nappes Helvétiques en Suisse orientale. Géobios Mém Spec, 3: 159-168

Funk H, Wyssling G (1987) Le Crétacé du shelf Helvétique - Interprétation des phases de "deepening" et de "shallowing". In: Salomon J (ed) Transgressions et régressions au Crétacé. Mém Géol Univ Dijon, 11: 145-147

Ganz E (1912) Stratigraphie der mittleren Kreide (Gargasien, Albien) der oberen helvetischen Decken in den nördlichen Schweizeralpen. Neue Denkschr Schweiz Natf Ges, 42/1, pp 148

Gebhard G (1983) Stratigraphische Kondensation am Beispiel mittelkretazischer Vorkommen im perialpinen Raum. PhD Thesis Univ Tübingen, pp 145

Gebhard G (1985) Kondensiertes Apt und Alb im Helvetikum (Allgäu und Vorarlberg), Biostratigraphie und Fauneninhalt. In: Kollmann HA (ed) Beiträge zur Stratigraphie und Paläogeographie der mittleren Kreide Zentral-Europas. Österr Akad Wiss Schriftenr Erdwiss Komm, 7: 271-284

Glangeaud L, D'Albissin M (1958) Les phases tectoniques du NE du Dévoluy et leur influence structurologique. Soc Géol France Bull, 8: 675-688

Goldhammer RK, Dunn PA, Hardie LA (1987) High frequency glacio-eustatic sea level oscillations with Milankovitch characteristics recorded in middle Triassic platform carbonates in northern Italy. Am J Sci, 287: 853-892

Grant J, Bathmann UV (1987) Swept away: resuspension of bacterial mats regulates benthic-pelagic exchange of sulfur. Science, 236: 1472-1474

Greber EA (1987) Die Geologie der mittleren Churfirsten: unter besonderer Berücksichtigung von mittelkretazischen Spalten an der Grenze Schrattenkalk-/Garschella-Formation. Diploma Thesis, ETH Zürich, pp 142

Greber EA, Ouwehand PJ (1988) Spaltenfüllungen im Dach der Schrattenkalk-Formation. Eclogae Geol Helv, 81/2: 373-385

Gross TF, Williams AJ, Nowell ARM (1988) A deep-sea sediment transport storm. Nature (London), 331: 518-521

Haldimann PA (1977) Sedimentologische Entwicklung der Schichten an einer Zyklengrenze der Helvetischen Unterkreide. Mitt Geol Inst ETH Univ Zürich NF, 219, pp 184

Hallam A (1984) Pre-Quaternary sea level changes. Annu Rev Earth Planet Sci, 12: 205-243

Hallock P (1987) Fluctuations in the trophic resource continuum: a factor in global diversity cycles? Paleoceanography, 2/5: 457-471

Hallock P, Schlager W (1986) Nutrient excess and the demise of coral reefs and carbonate platforms. Palaios, 1: 389-398

Hamilton D, Sommerville JH, Stanford PM (1980) Bottom currents and shelf sediments, southwest of Britain. J Sediment Pet, 26: 115-138

Hancock JM, Kauffman EG (1979) The great transgressions of the late Cretaceous. J Geol Soc Lond, 136: 175-186

Haq UB, Hardenbol J, Vail PR (1987) Chronology of fluctuating sea levels since the Triassic. Science, 235: 1156-1167

Harland WB, Cox AV, Llewellyn PG, Pickton CAG, Smith AG, Walters R (1982) A geological time scale. Cambridge Univ Press, pp 131

Harris PM, Frost SH, Seiglie GA, Schneidermann N (1984) Regional unconformities and depositional cycles, Cretaceous of Arabian peninsula. In: Schlee JS (ed) Inter-regional unconformities and hydrocarbon accumulations. Am Assoc Pet Geol Mem, 36: 67-80

Heggie DT, O'Brien GW, Reimers C, Burnett WC, McArthur JM, Blanks A, Skyring GW, Herczeg A, Milnes AR, Riggs SR, Cook PJ, Moriarty D, Hill PJ (in press) Iron-phosphorus cycling, sediment mixing, and the formation of modern marine phosphorites on the east Australian continental margins. Nature (London)

Heim A (1905) Der westliche Teil des Säntisgebirges. Beitr Geol Karte Schweiz NF, 16/2: 313-517

Heim A (1908) Über rezente und fossile subaquatische Rutschungen und deren lithologische Bedeutung. Neues Jahrb Miner Geol Paläontol, 2: 136-157

Heim A (1909) Sur les zones paléontologiques et lithologiques du Crétacique moyen dans les Alpes suisses. Bull Soc Géol France, 4/9: 101-127

Heim A (1910-17) Monographie der Churfirsten-Mattstock-Gruppe. Beitr Geol Karte Schweiz NF, 20/1: 1-272 (1910); 20/2: 273-368 (1913); 20/3: 369-573 (1916); 20/4: 573-662 (1917)

Heim A (1919) Zur Geologie des Grünten im Allgäu. Vjschr Natf Ges Zürich, 64: 458-486

Heim A (1923) Beobachtungen in den Vorarlberger Kreideketten. Eclogae Geol Helv, 18/2: 207-211

Heim A (1924) Über submarine Denudation und chemische Sedimente. Geol Rundschau, 15/1: 1-47

Heim A (1934) Stratigraphische Kondensation. Eclogae Geol Helv, 27: 372-383

Heim A (1946) Problemas de erosión submarina y sedimentación pelágica del presente y del pasado. Rev Mus La Plata Secc Geol, 4: 125-178

Heim A (1958) Oceanic sedimentation and submarine discontinuities. Eclogae Geol Helv, 51/3: 642-648

Heim A, Baumberger, E (1933) Jura und Kreide in den helvetischen Alpen beiderseits des Rheins (Vorarlberg und Ostschweiz). Denkschr Schweiz Natf Ges, 68/2, pp 184

Heim A, Seitz O (1934) Die mittlere Kreide in den helvetischen Alpen von Rheintal und Vorarlberg und das Problem der Kondensation. Denkschr Schweiz Natf Ges, 69/2: 185-310

Hickey LJ (1984). Changes in the angiosperm flora across the Cretaceous-Tertiary boundary. In: Berggren WA, Van Couvering JA (eds) Catastrophes and earth history. Princeton Univ. Press, pp 279-313

Hsü KJ (1982) Geosynclines in plate-tectonic settings: sediments in mountains. In: Hsü KJ (ed) Mountain building processes. Academic Press, London New York, pp 3-12

Iijima A, Hein JR, Siever R (1983) An introduction to siliceous deposits in the Pacific region. In: Iijima A, Hein JR, Siever R (eds) Siliceous deposits in the

Pacific region. Dev Sediment, Elsevier, Amsterdam, 36: 1-16

Jacob C, Tobler A (1906) Etude stratigraphique et paléontologique du Gault de la vallée de la Engelberger Aa (Alpes calcaires suisses, environs du Lac des Quatre Cantons). Mém Soc Paléont Suisse, 33: 3-26

Jahnke RA, Emerson SR, Roe KK, Burnett WC (1983) The present day formation of apatite in Mexican continental margin sediments. Geochim Cosmochim Acta, 47: 259-266

Jansson M (1986) Nitrate as a catalyst for phosphorus mobilization in sediments. In: Sly PG (ed) Sediments and water interactions. Springer, Berlin Heidelberg New York Tokyo, pp387-389

Jeletzky JA (1977) Causes of Cretaceous oscillations of sea level in western and Arctic Canada and some general geotectonic implications. Palaeontol Soc Japan Spec Pap, 21: 233-246

Joseph P, Beaudoin B, Cabrol C, Fries G (1986) Tectonics or differential compaction? The example of Banon fault trough (Apto-Albian, France S.E.). Int Assoc Sediment Canberra, (abstr) p 158

Juignet P, Rioult M, Destombes P (1973) Boreal influences in the Upper Aptian-Lower Albian beds of Normandy, northwest France. In: Casey R, Rawson PF, The Boreal Lower Cretaceous. Geol J Spec Issues, 5: 303-326

Kauffman EG (1984) The fabric of Cretaceous marine extinctions. In: Berggren WA, Van Couvering (eds) Catastrophes and earth history. Princeton Univ Press, pp 151-246

Keller B (1983) Geologie des Niderbauen, unter besonderer Berücksichtigung des Schrattenkalks, des Gault und des Seewerkalks. Diploma Thesis, Univ Zürich, pp 118

Kemper E (1982) Die Aucellinen des Apt und Unter-Alb Nordwestdeutschlands. Geol Jahrb Ser A, 65: 655-680

Kemper E (1983) Über Kalt- und Warmzeiten der Unterkreide. Zitteliana, 10: 359-369

Kemper E (1987) Das Klima der Kreide-Zeit (Teil 1). Geol Jahrb Ser A, 96: 5-185

Kendall GSC, Schlager W (1981) Carbonates and relative changes in sea level. Mar Geol, 44: 181-212

Kennedy WJ, Garrison RE (1975) Morphology and genesis of nodular phosphates in the Cenomanian glauconite marl of south east England. Lethaia, 8: 339-360

Kennett J (1982) Marine Geology. Prentice-Hall, New Jersey, pp 813

Korner M (1978) Sedimentologisch-stratigraphische Untersuchungen im helvetischen Gault zwischen Aare und Linth. PhD Thesis, Univ Bern, pp 136

Krajewski KP (1984) Early diagenetic phosphate cements in the Albian condensed glauconitic limestone of the Tatra Mountains, western Carpathians. Sedimentology, 31: 443-470

Krumbein EW (1983) Biogene Lamination, Stromatolith und Biostrom. Rutle-Festschr, Wellenburger Akad, pp 133-142

Kuenen PH (1958) Problems concerning source and transportation of flysch sediments. Geol Mijnbouw, 20: 329-339

Lamboy ML, Monty CLV (1987) Bacterial origin of phosphatized grains. Terra Cogn, (abstr) 7/2-3, p 207

Laubscher H, Bernoulli D (1982) History and deformation of the Alps. In: Hsü KJ (ed) Mountain building processes. Academic Press, London New York, pp 169-180

Le Pichon X, Bergerat F, Roulet MJ (1988) Plate kinematics and tectonics leading to the Alpine belt formation; a new analysis. In: Clark SP, Burchfiel, BC, Suppe J (eds) Processes in continental lithospheric deformation. Geol Soc Am Spec Pap, 218: 111-131

Lienert O (1965) Stratigraphie der Drusbergschichten und des Schrattenkalks unter besonderer Berücksichtigung der Orbitoliniden. Mitt Geol Inst Univ ETH Zürich NF, 65, pp 141

Lucas J, Prévôt L (1984) Synthèse de l'apatite par voie bactérienne à partir de matière organique et de divers carbonates de calcium dans des eaux douce et marine naturelles. Chem Geol, 42: 101-118

Lucas J, Prévôt L (1985) The synthesis of apatite by bacterial activity: mechanisms. Sci Géol Mém, 77: 83-92

Luyendyk BP, Forsyth D, Phillips JD (1972) Experimental approach to the paleocirculation of the oceanic surface waters. Geol Soc Am Bull, 83: 2649-2664

Manheim F, Rave G, Jipa D (1975) Marine phosphorite formation off Peru. J Sediment Pet, 45/1: 243-251

Matsumoto T (1980) Inter-regional correlation of transgressions and regressions in the Cretaceous period. Cret Res, 1/4: 359-373

Mikhailova IA (1979) The evolution of Aptian ammonoids. Paleontol J, 13/3: 267-274

Moslow TF, Nummedal D, Coleman JM (1988) Clastics. Geotimes, 33/2: 11-12

Müller PJ, Suess E (1979) Productivity, sedimentation rate, and sedimentary organic matter in the oceans - I. Organic carbon preservation. Deep-Sea Res, 26A: 1347-1362

Mullins HT, Rasch RF (1985) Sea-floor phosphorites along the central California continental margin. Econ Geol, 80: 696-715

Mullins HT, Thompson JB, McDougall K, Vercoutere TL (1985) Oxygen-minimum zone edge effects: evidence from the central California coastal upwelling system. Geology, 13: 491-494

Mutterlose J (1987) Calcareous nannofossils and belemnites as warm water indicators from the NW-German Middle Aptian. In: Kemper E (ed) Das Klima der Kreide-Zeit. Geol Jahrb Ser A, 96: 293-313

Naidin DP, Sasonova IG, Pojarkova ZN, Djalilov MR, Papulov GN, Senkovsky YN, Benjamovsky VN, Kopaevich LF (1980) Cretaceous transgressions and regressions on the Russian platform, in Crimea and central Asia. Cret Res, 1/4: 375-384

Oberhänsli H (1978) Mikropaläontologische und sedimentäre Untersuchungen in der Amdener Formation. Beitr Geol Karte Schweiz NF, 150, pp 83

Oberhauser R (1953) Geologische Untersuchungen im Flysch und Ultrahelvetikum der Hohen Kugel (Vorarlberg). Verh Geol Bundesanst Vienna, pp 176-183

Oberhauser R (1958) Neue Beiträge zur Geologie und Mikropaläontologie von Helvetikum und Flysch im Gebiet der Hohen Kugel (Vorarlberg). Verh Geol Bundesanst Vienna, pp 121-140

Oberhauser R (1973) Stratigraphisch-paläontologische Hinweise zum Ablauf tektonischer Ereignisse in den Ostalpen während der Kreidezeit. Geol Rundschau, 62: 96-106

Oberhauser R (1982) Geologische Karte der Republik Österreich, Bl. 100, St. Gallen und 111, Dornbirn, 1:25'000. Geol Bundesanst Vienna

O'Brien GW (1986) Reworking: a major control on the uranium-series and major element chemistry of phosphorites from the east Australian continental margin. Int Assoc Sediment, Canberra, (abstr) p 231

O'Brien GW, Harris JR, Milnes AR, Veeh HH (1981) Bacterial origin of east Australian continental margin phosphorites. Nature (London), 288: 442-444

O'Brien GW, Heggie D (1988) East Australian continental margin phosphorites. EOS, 69/1: 2

Ouwehand PJ (1986) Werden Phosphorite wirklich frühdiagenetisch gebildet? Beispiele aus der helvetischen Mittelkreide der Ostschweiz. In: Bechstädt T, Knitter M (eds) Symposiumband 1. Treffen deutschspr Sediment, Freiburg, pp 86-89

Ouwehand PJ (1987) Die Garschella-Formation ("Helvetischer Gault", Aptian-Cenomanian) der Churfirsten-Alvier Region (Ostschweiz). Sedimentologie, Phosphoritgenese, Stratigraphie. PhD Thesis, ETH Zürich, pp 296

Ouwehand PJ, Föllmi KB (1987) Mid-Cretaceous sedimentary sequences in the Alpine Helvetic realm of eastern Switzerland and Vorarlberg (Austria). 3rd Int Cret Symp, Tübingen, (abstr) p 44

Owen HG (1971) Middle Albian stratigraphy in the Anglo-Paris Basin. Bull Brit Mus Nat Hist Geol Suppl, 8, pp 1-164

Owen HG (1975) The stratigraphy of the Gault and Upper Greensand of the Weald. Proc Geol Assoc, 86: 475-498

Pantic N, Gansser A (1977) Palynologische Untersuchungen in Bündnerschiefern. Eclogae Geol Helv, 70/1: 59-81

Pantic N, Isler A (1978) Palynologische Untersuchungen in Bündnerschiefern (II). Eclogae Geol Helv, 71/3: 447-465

Parrish JT (1982) Upwelling and petroleum source beds, with reference to the Paleozoic. Am Assoc Pet Geol Bull, 66: 750-774

Parrish JT, Curtis RL (1982) Atmospheric circulation, upwelling, and organic-rich rocks in the Mesozoic and Cenozoic eras. Palaeogeogr Palaeoclimatol Palaeoecol, 40: 31-66

Pedley HM, Bennett SM (1985) Phosphorites, hardgrounds and syndepositional solution subsidence: a palaeoenvironmental model from the Miocene of the Maltese islands. Sediment Geol, 45: 1-34

130

Pindell J, Dewey JF (1982) Permo-Triassic reconstruction of western Pangea and the evolution of the Gulf of Mexico/Caribbean region. Tectonics, 1/2: 179-211

Pinet PR, Popenoe P (1985) A scenario of Mesozoic-Cenozoic ocean circulation over the Blake-Plateau and its environs. Geol Soc Am Bull, 96: 618-626

Pisciotto KA, Garrison RE (1981) Lithofacies and depositional environments of the Monterey Formation, California. In: Garrison RE, Douglas RG (eds) The Monterey Formation and related siliceous rocks of California. Soc Econ Paleontol Mineral Pacif Sect, 15: 97-122

Probst P (1980) Die Bündnerschiefer des nördlichen Penninikums zwischen Valser Tal und Passo di San Giacomo. Beitr Geol Karte Schweiz NF, 153, pp 63

Rampino MR, Stothers RB (1987) Episodic nature of the Cenozoic marine record. Paleoceanography, 2/3: 255-258

Raup DM, Sepkoski JJ Jr (1986) Periodic extinction of families and genera. Science, 231: 833-836

Read JF (1985) Carbonate platform facies models. Am Assoc Pet Geol Bull, 69/1: 1-21

Reimers CE, Kastner M, Garrison RE (in press) The role of bacterial mats in phosphate mineralization with particular reference to the Monterey Formation. In: Burnett WC, Riggs SR (eds) Genesis of Neogene to modern phosphorites. Cambridge Univ Press

Reimers CE, Suess E (1983) Spatial and temporal patterns of organic matter accumulation on the Peru continental margin. In: Thiede J, Suess E (eds) Coastal upwelling, its sediment record. Plenum Press, New York, B, pp 311-346

Reyment RA, Bengston P (1981) Aspects of Mid-Cretaceous regional geology. Academic Press, London New York, pp 327

Reyment RA, Bengston P (1985) Mid-Cretaceous events. Report on results 1974-1983. Publ Palaeontol Inst Univ Uppsala Spec Vol, 5; 1-132

Reyment RA, Bengston P (1986) Events of the Mid-Cretaceous. Phys Chem Earth, Pergamon Press, Oxford, 16: 1-209

Rich PV, Rich TH, Wagstaff BE, McEwen-Mason J, Douthitt CB, Gregory RT, Felton EA (1988) Evidence for low temperatures and biologic diversity in Cretaceous high latitudes of Australia. Science, 242: 1403-1406

Rick B (1985) Geologie der Flubrig unter besonderer Berücksichtigung der Altmann-Schichten und des "Gault". Diploma Thesis, ETH Zürich, pp 83

Ricou LE, Siddans AWB (1986) Collision tectonics in the western Alps. In: Coward MP, Ries AC (eds) Collision tectonics. Geol Soc Lond Spec Publ, 19: 229-244

Riggs SR (1984) Paleoceanographic model of Neogene phosphorite deposition, U.S. Atlantic continental margin. Science, 223: 123-131

Robaszynski F, Caron M (1979) Atlas de foraminifères planctoniques du Crétacé moyen (Mer Boreale et Téthys). Cah Micropaléont, 1: 1-185; 2: 1-181

Ross CA, Ross JRP (1985) Late Paleozoic depositional sequences are synchronous and worldwide. Geology, 13: 194-197

Savostin LA, Sibuet JC, Zonenshain LP, Le Pichon X, Roulet MJ (1986) Kinematic evolution of the Tethys belt from the Atlantic Ocean to the Pamirs since the Triassic. Tectonophysics, 123: 1-35

Schaub HP (1936) Geologie des Rawilgebietes. Eclogae Geol Helv, 29/2: 337-407

Schaub HP (1948) Über Aufarbeitung und Kondensation. Eclogae Geol Helv, 41/1: 89-94

Schlager W (1981) The paradox of drowned reefs and carbonate platforms. Geol Soc Am Bull, 92: 197-211

Schlanger SO (1986) High frequency sea level fluctuations in Cretaceous times: an emerging geophysical problem. In: Hsü KJ (ed) Mesozoic and Cenozoic oceans. Am Geophys Union, Geodyn Ser, 15: 61-74

Schlanger SO, Arthur MA, Jenkyns HC, Scholle PA (1987). The Cenomanian-Turonian oceanic anoxic event. I. Stratigraphy and distribution of organic carbon-rich beds and the marine $\delta^{13}C$ excursion. In: Brooks J, Fleet AJ (eds) Marine petroleum source rocks. Geol Soc Lond Spec Publ, 26: 371-401

Schlanger SO, Jenkyns HC (1976) Cretaceous oceanic anoxic events: causes and consequences. Geol Mijnbouw, 55: 179-184

Schlanger SO, Jenkyns HC, Premoli-Silva I (1981) Volcanism and vertical tectonics in the Pacific basin related to global Cretaceous transgressions. Earth Planet Sci Lett, 52: 435-449

Schneider SH, Thompson SL, Barron EJ (1985) Mid-Cretaceous continental surface temperatures: are high CO_2 concentrations needed to simulate above-freezing winter conditions. In: Sundquist ET, Broecker WS (eds) The carbon cycle and atmospheric CO_2: natural variations Archean to present. Am Geophys Union, Geophys Monogr, 32: 554-559

Scholle PA, Arthur MA (1980) Carbon isotope fluctuations in Cretaceous pelagic limestones: potential stratigraphic and petroleum exploration tool. Am Assoc Pet Geol Bull, 64/1: 67-87

Scholz HH (1979) Paläontologie, Aufbau und Verbreitung der Bioherme und Biostrome im Allgäuer Schrattenkalk (Helvetikum, Unterkreide). PhD Thesis, Univ München, pp 133

Scholz HH (1984) Sklerospongia aus dem Allgäuer Schrattenkalk (Helvetikum, Bayerische Alpen). Neues Jahrb Geol Paläontol Mh, pp 645-653

Schumann EH, Li Van Heerden I (1988) Observations of Agulhas current frontal features south of Africa, October 1983. Deep-Sea Res, 35/8: 1355-1362

Schwan W (1980) Geodynamic peaks in alpinotyp orogenies and changes in ocean-floor spreading during late Jurassic-late Tertiary time. Am Assoc Pet Geol Bull, 64/3: 359-373

Seidov DG (1986) Numerical modelling of the ocean circulation and paleo-circulation. In: Hsü KJ (ed) Mesozoic and Cenozoic oceans. Am Geophys Union, Geodyn Ser, 15: 11-26

Seilacher A, Reif WE, Westphal F (1985) Sedimentological, ecological, and temporal patterns of fossil Lagerstätten. Philos Trans R Soc Lond Ser B, 311: 5-23

Seiglie GA, Baker MB (1984) Relative sea level changes during the middle and late Cretaceous from Zaire to Cameroon (central west Africa). In: Schlee JS (ed) Interregional unconformities and hydrocarbon accumulation. Am Assoc Pet Geol Mem, 36: 7-36

Sepkoski JJ Jr (1989) Periodicity in extinction and the problem of catastrophism in the history of life. J Geol Soc Lond, 146: 7-19

Shaffer G (1986) Phosphate pumps and shuttles in the Black Sea. Nature (London), 321: 515-517

Shanmugam G (1988) Origin, recognition, and importance of erosional unconformities in sedimentary basins. In: Kleinspehn KL, Paola C (eds) New perspectives in basin analysis. Springer, Berlin Heidelberg New York Tokyo, pp 83-108

Slansky M (1986) Geology of sedimentary phosphates. North Oxford Academic, London, pp 210

Smith AG, Woodcock NH (1982) Tectonic syntheses of the Alpine-Mediterranean region: a review. In: Hsü KJ (ed) Alpine-Mediterranean geodynamics. Am Geophys Union, Geodyn Ser 7: 15-38

Smith DB (1988) Bypassing of sand over sand waves and through a sand wave field in the central region of the southern North Sea. In: De Boer PL, Van Gelder A, Nio SD (eds) Tide-influenced sedimentary environments and facies. Reidel, Dordrecht, pp 39-50

Soudry D (1987) Ultra-fine structures and genesis of the Campanian Negev high-grade phosphorites (southern Israel). Sedimentology, 34: 641-660

Soudry D, Champetier Y (1983) Microbial processes in the Negev phosphorites (southern Israel). Sedimentology, 30; 411-423

Soudry D, Lewy Z (1988) Microbially influenced formation of phosphate nodules and megafossil moulds (Negev, southern Israel). Palaeogeogr Palaeoclimatol Palaeoecol, 64: 15-34

Soutar A, Johnson SR, Baumgartner TR (1981) In search of modern depositional analogs to the Monterey Formation. In: Garrison RE, Douglas RG, Pisciotto KE, Isaacs CM, Ingle JC (eds) The Monterey Formation and related siliceous rocks of California. Soc Econ Paleontol Mineral Pacif Sect, 15: 123-147

Speyer SE, Brett CE (1988) Taphonomic models for epeiric sea environments: middle Paleozoic examples. Palaeogeogr Palaeoclimatol Palaeoecol, 63: 225-262

Spicer RA (1987) The significance of the Cretaceous flora of northern Alaska for the reconstruction of the climate of the Cretaceous. In: Kemper E (ed) Das Klima der Kreide-Zeit. Geol Jahrb Ser A, 96: 265-291

Spicher A (1980) Geologische Karte der Schweiz. Schweiz Geol Komm, Bern

Strasser A (1979) Betlis-Kalk und Diphyoideskalk (ca. Valanginian) in der Zentral- und Ostschweiz. Mitt Geol Inst ETH Univ Zürich NF, 225, pp 208

Strasser A (1982) Fazielle und sedimentologische Entwicklung des Betlis-Kalkes (Valanginian) im Helvetikum der Zentral- und Ostschweiz. Eclogae Geol Helv, 75/1: 1-22

Strasser A (1984) Black-pebble occurrence and genesis in Holocene carbonate sediments (Florida Keys, Bahamas, and Tunis). J Sediment Pet, 54/4: 1097-1109

Stride AH (1988) Preservation of marine sand wave structures. In: De Boer

PL, Van Gelder A, Nio SD (eds) Tide-influenced sedimentary environments and facies. Reidel, Dordrecht, pp 13-22

Stumm W, Leckie JO (1970) Phosphate exchange with sediments; its role in the productivity of surface waters. Adv Water Pollution Res, 2, III-26: 1-16

Suess E (1883) Das Anlitz der Erde, Teil 1. Tempsky, Prague, pp 778

Suess E (1981) Phosphate regeneration from sediments of the Peru continental margin by dissolution of fish debris. Geochim Cosmochim Acta, 45: 577-588

Summerhayes CP (1987) Organic-rich Cretaceous sediments from the north Atlantic. In: Brooks J, Fleet AJ (eds) Marine petroleum source rocks. Geol Soc Lond Spec Publ, 26: 301-316

Swift DJP, Rice DD (1984) Sand bodies on muddy shelves: a model for sedimentation in the Western Interior Seaway, North America. In: Tillman RW, Siemers CT (eds) Siliciclastic shelf sediments. Soc Econ Paleontol Mineral Spec Publ, 34: 43-62

Thierstein HR (1979) Paleoceanographic implications of organic carbon and carbonate distribution in Mesozoic deep sea sediments. In: Talwani M, Hay WW, Ryan WFB (eds) Deep drilling results in the Atlantic Ocean: continental margins and paleoenvironment. Am Geophys Union, Maurice Ewing Ser, 3: 249-274

Thompson JB, Mullins HT, Newton CR, Vercoutere TL (1985) Alternative biofacies model for dysaerobic communities. Lethaia, 18: 169-179

Trümpy R (1973) The timing of orogenic events in the central Alps. In: De Jong K, Scholten R (eds) Gravity and tectonics. Wiley, New York pp 229-251

Trümpy R (1980) Geology of Switzerland. Wepf, Basel, pp 104

Trümpy R (1982) Alpine paleogeography: a reappraisal. In: Hsü KJ (ed) Mountain building processes. Academic Press, London New York, pp 149-156

Trümpy R (1985) Die Plattentektonik und die Entstehung der Alpen. Njblatt Natf Ges Zürich, 129/5: 5-47

Tschudy RH (1984) Palynological evidence for change in continental floras at the Cretaceous Tertiary boundary. In: Berggren WA, Van Couvering JA (eds) Catastrophies and earth history. Princeton Univ Press, pp 315-337

Tucholke BE, Vogt PR (1979) Western north Atlantic; sedimentary evolution and aspects of tectonic history. Init Rep DSDP, 43: 791-825

Vail PR, Mitchum RM Jr, Todd RG, Widmier JM, Thompson S III, Sangree JB, Bubb JN, Hatleilid WG (1977) Seismic stratigraphy and global changes of sea level. In: Payton CE (ed) Seismic stratigraphy - applications to hydrocarbon exploration. Am Assoc Pet Geol Mem, 26: 49-212

Vail PR, Hardenbol J, Todd RG (1984) Jurassic unconformities, chronostratigraphy and sea level changes from seismic and biostratigraphy. In: Schlee JS (ed) Interregional unconformities and hydrocarbon accumulation. Am Assoc Pet Geol Mem, 36: 7-36

Vakhrameev U (1978) The climates of the northern hemisphere in the Cretaceous in the light of paleobotanical data. Paleontol J, 2: 143-154

Vandenberg J (1979) Reconstructions of the western Mediterranean area for the Mesozoic and Tertiary time span. Geol Mijnbouw, 58/2: 153-160

Vercoutere TL, Mullins HT, McDougall K, Thompson JB (1987) Sedimentation across the central California oxygen minimum zone: an alternative coastal upwelling sequence. J Sediment Pet, 57/4: 709-722

Voigt E (1962) Frühdiagenetische Deformation der turonen Plänerkalke bei Halle/Westf. als Folge einer Grossgleitung unter besonderer Berücksichtigung des Phacoid-Problems. Mitt Geol Staatsinst Hamburg, 31: 146-275

Volk T (1987) Feedbacks between weathering and atmospheric CO_2 over the last 100 million years. Am J Sci, 287: 763-779

Weidich KF (1987) Das Ultrahelvetikum von Liebenstein (Allgäu) und seine Foraminiferenfauna. Zitteliana, 15: 193-217

Weimer RJ (1984) Relation of unconformities, tectonics and sea level changes, Cretaceous of Western Interior, USA. In: Schlee JS (ed) Interregional unconformities and hydrocarbon accumulation. Am Assoc Pet Geol Mem, 36: 7-36

Williams LA (1984) Subtidal stromatolites in the Monterey Formation and other organic rich rocks as suggested source contributors to petroleum formation. Am Assoc Pet Geol Bull, 68: 1879-1893

Williams LA, Reimers CE (1983) Role of bacterial mats in oxygen-deficient marine basins and coastal upwelling regimes: preliminary report. Geology, 11: 267-269

Winkler W, Lüdin P (1986) Flysch and melange formations related to early Alpine subduction in the eastern Alps (Switzerland, Austria, Germany). Int Assoc Sediment, Canberra, (abstr) p 333

Wyssling G (1985) Palinspastische Abwicklung der helvetischen Decken von Vorarlberg und Allgäu. Jahrb Geol Bundesanst Vienna, 127/4: 701-706

Wyssling G (1986) Der frühkretazische helvetische Schelf im Vorarlberg und Allgäu. Stratigraphie, Sedimentologie und Paläogeographie. Jahrb Geol Bundesanst Vienna, 129/1: 161-265

Yingst JY, Rhoads DC (1980) The role of bioturbation in the enhancement of bacterial growth rates in marine sediments. In: Tenore KR, Coull BC (eds) Marine benthic dynamics. Univ South Carolina Press, pp 407-421

Zacher W (1973) Das Helvetikum zwischen Rhein und Iller (Allgäu-Vorarlberg): tektonische, paläogeographische und sedimentologische Untersuchungen. Geotekt Forsch, 44: 1-74

Ziegler PA (1982) Geological atlas of western and central Europe. Elsevier, Amsterdam, pp 136

Ziegler PA (1987) Late Cretaceous and Cenozoic intra-plate compressional deformations in the Alpine foreland - a geodynamic model. Tectonophysics, 137: 389-420

Ziegler PA (1988) Evolution of the Artic-North Atlantic and the western Tethys. Am Assoc Pet Geol Mem, 43, pp 198

Subject Index

The numbers of text figures are given in italic type; page numbers appear in roman type.

Section Localities

Coordinates and locality names used here correspond to the topographic map
"Hoher Freschen", 1:50.000, no. 228, Swiss Federal Topographic Survey, 3084
Wabern (cf. Föllmi 1986)

Fig. 4.

Section 1.	Feldkirch	762.900/234.150/460
Section 2.	Schellenberg, Nofels	762.080/235.750/460
Section 3.	Moos, Emmabach	770.470/244.970/1050
Section 4.	U. Wäldle Alp	774.900/245.180/1070
Section 5.	Gunzmoos, Nest Alps	777.520/246.920/1230
Section 6.	Laubach Alp	778.050/247.300/1280
Section 7.	Hinterwang	773.030/240.250/1330
Section 8.	Mittagspitz	785.000/242.700/1900

Fig. 5.

A.	Unt. Wäldle Alp	774.900/245.180/1070
B.	Schellenberg, Nofels	762.080/235.750/460

Fig. 7.

Section 1.	Rappenloch bridge	776.820/250.820/610
Section 2.	Niedere, H. Knopf	776.500/249.100/810
Section 3.	N Götzis	768.000/245.900/460
Section 4.	Gsohl Alp	771.000/247.050/810
Section 5.	Örfla Gulch	767.850/244.400/525
Section 6.	Moos, Emmabach	770.470/244.970/1050
Section 7.	Gunzmoos Alp	778.000/248.300/970
Section 8.	Obersehren Alp	781.150/248.430/1450
Section 9.	Obersehren Alp	780.650/248.270/1510
Section 10.	Laubach Alp	778.050/247.300/1280
Section 11.	Ilgenwald Alp	775.300/245.550/1080
Section 12.	Hoher Freschen	777.230/242.070/1990
Section 13.	S Hoher Freschen	777.600/241.620/1900
Section 14.	Sünser Kopf	781.900/241.850/2010
Section 15.	Sünser Alp	781.420/240.580/1765
Section 16.	Street to Damuls	798.250/242.200/1080
Section 17.	Mellenbach	780.500/240.150/1560
Section 18.	Hohe Kugel	771.800/244.820/1530

Fig. 8.

	Niedere, H. Knopf	776.500/249.100/810

Fig. 9.

	N Götzis	768.000/245.900/460

150

Fig. 12 and 13.

| | Mellenbach | 780.500/240.150/1560 |

Fig. 14.

| A. | Staffel Alp | 771.870/244.250/1420 |
| B. | Hohe Kugel | 771.800/244.820/1530 |

Fig. 15.

Section 1.	Chäserrugg	742.650/224.750/2040
Section 2.	Stein	764.300/233.400/520
Section 3.	Strahlkopf	772.250/247.700/950
Section 4.	Langensack	779.650/250.670/1100
Section 5.	Bocksberg	775.050/248.370/1160
Section 6.	Hohe Lug, Emmabach	769.370/244.600/830
Section 7.	Tschütsch	765.530/242.320/450
Section 8.	Klaus	766.950/242.470/500
Section 9.	Simonsbach	783.750/248.400/780
Section 10.	Rankweil	767.950/238.440/500

Fig. 18.

| A. | Gsohl Alp | 770.620/247.070/700 |
| B. | Klaus | 767.600/242.750/590 |

Fig. 19.

Section 1.	Fallbach	774.500/250.250/890
Section 2.	Klausberg	781.800/252.850/670
Section 3.	Hof	784.450/250.950/640
Section 4.	Müselbach	777.750/250.650/730
Section 5.	Alploch Gulch	776.500/250.050/600
Section 6.	Breiterberg	772.800/249.550/890
Section 7.	Schwarzenberg	773.050/248.950/1040
Section 8.	N Götzis	768.000/245.900/460
Section 9.	Gsohl Alp	771.000/247.050/810
Section 10.	Hohe Lug, Emmabach	769.370/244.600/830
Section 11.	Osanken	768.020/243.070/640
Section 12.	Ache	777.300/247.900/980
Section 13.	Simonsbach	783.750/248.400/780
Section 14.	Dafins	770.250/240.800/750
Section 15.	Körb Alp	777.100/243.630/1510
Section 16.	Hohe Matona	777.750/241.030/1960
Section 17.	Street to Damuls	798.250/242.200/1080

Fig. 21.
From left to right

Finsternaubach	770.750/247.650/550
Hof	784.450/250.950/640
Müselbach	777.750/250.650/730

	N Götzis	768.000/245.900/460
	Strahlkopf	772.250/247.700/950
Fig. 22.		
	Steinriesler Bach	782.100/252.800/590
Fig. 23.		
	Schwarzenberg	773.050/248.950/1040
Fig. 24. From left to right		
	Breiterberg	772.800/249.550/890
	Rappenloch bridge	776.820/250.820/610
	Örfla Gulch	767.850/244.400/525
	Hohe Lug, Emmabach	769.370/244.600/830
	Klaus	766.950/242.470/500
	Ache	777.300/247.900/980
	Stein	764.300/233.400/520
Fig. 25. A. B.	Steinriesler Bach Osanken	782.100/252.800/590 768.020/243.070/640
Fig. 26.		
	V. Schaner Alp	776.470/247.320/980
Fig. 27.		
	Strahlkopf	772.430/247.270/1350
Fig. 29. Section 1. Section 2. Section 3. Section 4. Section 5. Section 6. Section 7.	Müselbach Unterklien Sch. Mann Alp Simonsbach U. Saluver Alp SW Hohe Matona Hohe Matona	777.750/250.650/730 772.950/250.650/450 772.900/247.150/1300 783.750/248.400/780 776.850/239.870/1570 777.180/240.350/1820 777.680/241.000/1990
Fig. 31.		
	Müselbach	777.750/250.650/730
Fig. 32. Section 1. Section 2. Section 3.	Örfla Gulch Hoher Freschen S Hoher Freschen	767.850/244.400/525 777.230/242.070/1990 777.600/241.620/1900

Fig. 33.
A.   Götznerberg   768.470/246.000/480
B.   Emserhalde   768.900/246.200/550

Fig. 34.
Section 1.   Finsternaubach   770.750/247.600/570
Section 2.   Finsternaubach   771.070/247.800/660
Section 3.   Fallbach   774.500/250.250/890
Section 4.   Klausberg   781.800/252.850/670
Section 5.   Alploch Gulch   776.500/250.050/600
Section 6.   E Gütle   777.350/251.120/600
Section 7.   Müselbach   777.750/250.650/730
Section 8.   Öfen   779.000/251.000/1020
Section 9.   Simonsbach   783.750/248.400/800
Section 10.   Säck Alp   779.030/250.350/1020
Section 11.   Bocksberg   774.850/248.330/1180
Section 12.   Rudachbach   777.600/249.030/870
Section 13.   Malertobel   771.370/240.920/960
Section 14.   Rotwald   771.750/240.900/1230
Section 15.   Hasengerach Alp   779.200/248.050/1340
Section 16.   First   772.800/244.380/1590
Section 17.   Ratzbach   770.950/243.400/1010
Section 18.   Röthis   767.750/241.050/580
Section 19.   Röfix, Röthis   768.800/240.700/580
Section 20.   Körb Alp   777.100/243.580/1520

Fig. 35.
                      Rudachbach   777.600/249.030/870

Fig. 37.
A.   Kobel Ache   777.470/250.020/790
B.   Finsternaubach   770.750/247.600/570

Fig. 38.
                      S Hof   783.850/250.650/680

Fig. 39.
                      Gsohl Alp   770.850/247.050/710

Fig. 40.
                      Hasengerach Alp   779.420/248.250/1260

Fig. 41.
                      Staffel Alp   772.200/244.450/1440

Fig. 42.
                      Rudachbach   777.550/249.900/790

Fig. 45.

A. E Gütle 777.350/251.120/600

B. Steinriesler Bach 782.100/252.800/590

Fig. 46.

Simonsbach 783.750/248.400/780

Fig. 47.

Bezau ?

Fig. 48 and 52.

Unterklien 772.050/250.300/460

Fig. 53.

Kühberg Alp 774.650/250.850/910